T0401271

precisely and immediately imparted to the surgeon's hands. We would need to program the information on variable senses of touch to this system to create a database. In Chapter 1, by assessing the efficiency of wound healing in terms of the extent of angiogenesis and the expressions of growth factors at wound sites, it was determined that a knot-tying force of 1.5 N may be the most appropriate for optimal wound healing in the gastrointestinal tract. We believe that the results of our study provide useful data for surgery and will contribute to the future development of robotic forceps with tactile sensation. This technology can be applied not only to robotic surgery or telesurgery but also to endoscopic methods and interventional radiology.

Inspired by the organized behaviors of honeybee swarms, an individual-based mathematical model is proposed in Chapter 2 for the heterogeneous swarm. The heterogeneous swarm is assumed to consist of two different kinds of individuals, namely, the scouts and the normal agents, with respect to their sensing abilities. Besides, a short-distance-bounded-attraction function was proposed to describe the attraction among individuals.

Firstly the heterogeneous swarm model is identified and the swarm cohesion is proved, and the analytical bound on the swarm size is provided. Secondly, the foraging properties of the heterogeneous swarm in multimodal Gaussian environment are studied, and conditions for collective convergence to more favorable regions are provided. Thirdly, simulations were carried out and the priority of proposed short-distance-bounded-attraction function was demonstrated in complex environment. Simulation results show that the heterogeneous swarm model provides a feasible framework for multi-robot navigation applications.

In Chapter 3, in order to study autonomous behaviors in small military robots, researchers at the U.S. Army Research Laboratory (ARL) renovated an existing but outdated ATRV research robot. Commercial sensors with capabilities resembling those anticipated from the military tech base were selected and integrated, and the computing capability was substantially enhanced by judiciously selecting commercial components. Support electronics were upgraded or replaced as necessary. Safety elements common in larger robotic vehicles were integrated into the small ATRV chassis. Systems software was selected to provide a stable foundation for the advanced functions envisioned. Player, a middleware widely used by academic robot researchers, was incorporated as a springboard to the agent-based behaviors believed necessary for the next phase of development in robotics. A distributed development environment was implemented to enable parallel software efforts. Issues in software architecture were identified, and architectures from the literature were investigated in search of a foundation for future work. Without major investment, the antiquated research robot has become

PREFACE

Up to now, industrial robots have drastically rationalized many kinds of manufacturing processes in industrial fields. The user interface provided by the robot maker has been almost limited to so-called teaching pendant - a useful and safe tool to obtain the position and orientation at the tip of a robot along a desired trajectory. This book examines a position/force control system, first designed for industrial robots with an open architecture controller. In addition, this book provides an analysis of various tracking techniques (optical, electromagnetic, ultrasonic and laser-based) in different environments and surgical fields. An overview of registration algorithms is also given, with an analysis of theoretical registration error-distribution followed by practical design considerations for an accurate registration set-up. The authors of this book also summarize the registration methods used for commercial and research surgical robots. Other chapters assess the development of a system of haptic forceps with bilateral control, allowing tactile sensation to be precisely and immediately imparted to the surgeon's hands. The authors believe that the results of the study will provide useful data for surgery and will contribute to the future development of robotic forceps with tactile sensation.

Tactile sensation of tissues is not imparted to the surgeon's hands from the tip of the forceps in robotic surgery, making the use of appropriate force for surgical techniques difficult. To overcome this drawback, studies on haptic forceps have already been conducted; forceps with an infrared sensor, a piezoelectric transducer, the PHANToM force feedback device, a pneumatic haptic feedback actuator array, and so on. Tactile sensation, which is different from other sensations, participates in both the characterization of subject matter and the force and location of the recipient, i.e. bilaterality. We are now planning to develop a system of haptic forceps with bilateral control allowing tactile sensation to be

CONTENTS

LIBRARY OF CONGRESS CATALOGING-IN-PUBLICATION DATA

New robotics research / editors, Ean D. Wagner and Lawrence G. Kovacs.
p. cm.
Includes index.
ISBN 978-1-60741-093-5 (hardcover)
1. Robotics--Research. I. Wagner, Ean D. II. Kovacs, Lawrence G.
TJ211.N52 2009
629.8'92--dc22

2009040791

Published by Nova Science Publishers, Inc. ✢ New York

NEW ROBOTICS RESEARCH

EAN D. WAGNER
AND
LAWRENCE G. KOVACS
EDITORS

Nova Science Publishers, Inc.
New York

NEW ROBOTICS RESEARCH

a key element in ARL's quest to develop technologies for a highly capable robot to team with soldiers on tomorrow's urban battlefield.

The project in Chapter 4 is closely tied with the ongoing work of visiting Professor Kenn Oldham and the U.S. Army Research Laboratory's (ARL) joint effort on creating highly flexible, large payload capacity joints for a ground mobile millimeter-scale robot. The fabrication process to add parylene coatings to the piezo-microelectromechanical systems (piezoMEMS) actuator process has been characterized using test structures. Scanning electron and optical microscopy of the joint assemblies; analysis of the coating technology for trench fill; process robustness to exposure to solvents and photolithographic processing; and adhesion of parylene to both platinum and lead zirconate titanate (PZT) thin films have been completed on two separate fabrication sequences. Parylene coatings have been successfully applied to both platinum and PZT thin films and the challenges associated with parylene survival with multiple fabrication process steps have been evaluated. Future work will include full release of test structures on the existing wafers in fabrication as well as implementation of process improvements into a fully functional piezoMEMS plus parylene actuator joint.

Recently we have presented a system for panoramic depth imaging with a single standard camera. The system is mosaic-based, which means that we use a single standard rotating camera and assemble the captured images in a multiperspective panoramic image. Due to a setoff of the camera's optical center from the rotational center of the system we are able to capture the motion parallax effect from a single sweep around the rotational center, which enables the stereo reconstruction. One of the problems of such a system is the fact that we cannot generate a stereo pair of images in real time. Chapter 5 presents a possible solution to this problem, which is based on simultaneously using many standard cameras. We perform simulations on real scene images to establish the quality of new sensor results in comparison to results obtained with the old sensor. The goal of the chapter is to reveal whether the new sensor can be used for real time capturing of panoramic depth images and consequently for autonomous navigation of a mobile robot in a room. In particular, we focus on the real time generation of panoramic stereo pairs since the calculation of depth images can already be run in real time. The basic panoramic depth imaging system and its real time extension are comprehensively analysed and compared. The analyses reveal a number of interesting properties of the systems. According to the basic system accuracy we definitely can use the system for autonomous robot localization and navigation tasks. The assumptions made in the real time extension of the basic system are proved to be correct, but the accuracy of the new sensor generally deteriorates in comparison to the basic sensor.

The ultimate objective of surgical robotic systems is to carry out a preoperative plan accurately with precision. A surgical robot is part of a larger system, comprised of imaging device(s), position tracking and various instruments. For effective and accurate cooperation between these systems, they must be registered to the patient.

The goal of intraoperative registration is to establish a relationship between the frame of reference of the robotic system and the preoperative plan, typically generated in the coordinate system of the imaging device. This step has crucial impact on the overall accuracy, since registration inaccuracies are significantly higher than those of the robots mechanical and control systems. The main causes of registration inaccuracy are limitations of the tracking system, including environmental interference, difficulty accessing anatomical landmarks, such as in minimally invasive surgery, and inherent inaccuracies of the registration algorithm, including improper selection of registration fiducials.

The first part of Chapter 6 provides an analysis of various tracking techniques (optical, electromagnetic, ultrasonic, and laser-based) in different environments and surgical fields. The second part of this chapter gives an overview of registration algorithms, with an analysis of theoretical registration error-distribution followed by practical design considerations for an accurate registration set-up. The final part of this chapter summarizes the registration methods used for commercial and research surgical robots.

In Chapter 7, a position/force control system is first designed for industrial robots with an open architecture controller. Position and orientation of the tool attached to the tip of an industrial robot are controlled based on the model designed by a CAD system. Also, force including kinetic friction is controlled through a desired impedance model. The both manipulated variables generated from the position control system and force control system are velocity quantity in Cartesian-coordinate system, so that the hybrid control system can be easily applied to industrial robots with an open architecture controller. Next, we introduce two examples of applications being utilized in actual manufacturing process. One is the 3D robot sander which sands the free-formed surface of wooden materials. The finished wooden workpiece with curved surface is used for a part constructing a piece of artistic furniture. The other is the mold polishing robot which finishes aluminum PET bottle blow molds. Further, the application limit of articulated-type industrial robots is quantitatively evaluated through a simple static position/force measurement. Finally, we consider a novel desktop orthogonal-type robot with higher position and force resolutions to finish a smaller workpiece such as a plastic lens mold which conventional articulated-type industrial robots have not been able to deal with. The basic position/force control performance is shown, and present research progress and promising future works are introduced.

In: New Robotics Research
Editors: E.D. Wagner et al, pp. 1-11

ISBN: 978-1-60741-093-5

Chapter 1

DEVELOPMENT OF HAPTIC FORCEPS FOR ROBOTIC SURGERY

Junya Oguma[1], Soji Ozawa[2,*], Yasuhide Morikawa[1], Yuko Kitagawa[1] and Kouhei Ohnishi[3]

[1]Department of Surgery, School of Medicine, Keio University, Japan
[2]Department of Gastroenterological Surgery, Tokai University School of Medicine, Japan
[3]Department of System Design Engineering, Keio University Faculty of Science and Technology, Japan

Abstract

Tactile sensation of tissues is not imparted to the surgeon's hands from the tip of the forceps in robotic surgery, making the use of appropriate force for surgical techniques difficult. To overcome this drawback, studies on haptic forceps have already been conducted; forceps with an infrared sensor, a piezoelectric transducer, the PHANToM force feedback device, a pneumatic haptic feedback actuator array, and so on. Tactile sensation, which is different from other sensations, participates in both the characterization of subject matter and the force and location of the recipient, i.e. bilaterality. We are now planning to develop a system of haptic forceps with bilateral control allowing tactile

* E-mail address: sozawa@tokai.ac.jp, fax: +81-463-95-6491, phone: +81-463-93-1121. Corresponding author information: Soji Ozawa, MD, PhD, FACS, Professor, Department of Gastroenterological Surgery, Tokai University School of Medicine, 143 Shimokasuya, Isehara, Kanagawa 259-1193, Japan.

sensation to be precisely and immediately imparted to the surgeon's hands. We would need to program the information on variable senses of touch to this system to create a database. In our study, by assessing the efficiency of wound healing in terms of the extent of angiogenesis and the expressions of growth factors at wound sites, it was determined that a knot-tying force of 1.5 N may be the most appropriate for optimal wound healing in the gastrointestinal tract. We believe that the results of our study provide useful data for surgery and will contribute to the future development of robotic forceps with tactile sensation. This technology can be applied not only to robotic surgery or telesurgery but also to endoscopic methods and interventional radiology.

Introduction

Endoscopic surgery has spread rapidly since the 1990 based on being minimally invasive surgery, and has now gained a firm footing as a standard method for some diseases. Endoscopic surgery is superior to open surgery from the viewpoints of an expanded operative view, decreased postoperative pain, reduction of hospitalization and early return to normal activities. On the other hand, there were clearly some problems in endoscopic surgery, such as the necessity for endoscopist, trembling and other hand movements of the endoscopist, the handling limitation due to lack of freedom at the tip of the forceps and so on [1,2,3]. Surgical robots were developed in several countries starting in the early 1990s for the purpose of performing solosurgery or telesurgery. As the tips of the forceps were more flexible owing to surgical robots, the drawbacks of endoscopic surgery were partly overcome, and the development and clinical application of these devices is now progressing in various fields as tools that will allow minimally invasive surgery. The daVinci that is an especially representative master-slave manipulator that has six axes of flexibility at the tip of forceps, can be controlled in three view dimensions, and has scaling and filtering functions. Thus, we can perform more precise surgical maneuvers that are very difficult with routine endoscopic surgery using the daVinci. On the other hand, robotic surgery clearly has several problems, e.g. the huge device, high cost, complicated setting and lack of force feedback to surgeons [4]. In the current robotic surgery, knot-tying or grasping force is decided based on visual information alone, such as the stretch of the thread and tissue deformation. The lack of force feedback is a problem that must be solved urgently for the spread of robotic surgery, because knot-tying and grasping force on weak tissue can influence the perioperative courses of patients. There has been some research on tactile sensation with robotic surgery, but none on clinical application. We review herein past reports on haptic forceps in robotic surgery, our basic research on the relation between appropriate knot-tying force and wound healing and, finally, future prospects.

Research of Force Feedback in Robotic Surgery

Medical robotics has developed in various fields [5]. The first reported robot applied to clinical treatment was the Robodoc in 1992, which was used for total hip arthroplasty [6]. Subsequently, AESOP [7], TISKA endoarm [8], Fips endoarm [9], Endoassist [10] and so on were reported as camera guidance systems. These were developed for solosurgery. On the other hand, a master-slave manipulator was developed for telesurgery. ARTEMIS, from Germany, was the first master-slave manipulator, but it was not used for clinical treatment [11]. The daVinci [12,13] and ZEUS [14] were then developed in the U.S. at the same time, and were applied to cardiovascular surgery, general surgery, urology and gynecology. More precise techniques would be possible with introduction of a surgical robot to endoscopic surgery, allowing the spread of robotic surgery expected. Several studies have compared the performance of robot-assisted and conventional surgeries [15,16]. It was suggested that lack of force feedback to surgeons in robotic surgery can cause complications such as accidental blood vessel puncture or tissue damage [17,18].

Bethea et al suggested that visual sensory substitution with a tension measuring device permitted the surgeon to apply more consistent, precise, and greater tension to fine suture materials without breakage during robot-assisted knot-tying [19]. This system allowed haptic information at the tip of the forceps to feedback to surgeons as visual information, but there are several problems with this system. It is necessary to obtain various haptic types of data on tissues, and when surgeons obtain visual information during an operation, they must look away from the operative field. There is a report on a sound feedback system [20], but interference from other sounds during the operation is a potential problem. Thorey et al suggested that force feedback added to vision feedback leading to an improvement in tissue characterization versus using only vision feedback or only force feedback [21]. Force feedback to surgeons is necessary in order to perform delicate and precise movements during an operation. The precision of a force feedback system is important for obtaining more accurate haptic information.

Various reports to date have described force feedback systems including piezoelectric [22], electromagnetic [23], servo motors [24],and pneumatic haptic feedback actuator array (Table 1) [25]. In fact, a few reports suggested force feedback systems might be programmed into surgical robots. PHANToM is an excellent force feedback system, and several reports have suggested this system for robotic surgery [21,26].

PHANToM, which was developed by Tomas Massie and Kenneth Salisbury of the Massachusetts Institute of Technology in 1993, is a force display, able to realize six degree-of-freedom input and three degree-of-freedom output [27]. It achieved an accurate sense of touch, previously unprecedented, and is now used worldwide. PHANToM estimates the movement of user's fingers and adds force to this point. Thus, it is able to produce haptic information which is very similar to a real sense of touch against objects with a certain hardness. It is now used most in research on haptics with robotic surgery. In recent years, it has become possible to use it on the desk via hardware miniaturization and is easily installed. On the other hand, the point at which the sense of touch is presented is limited to one per one machine. Thus, in order to present several points of sense of touch, the machine must be large-scale. Furthermore, the control space is narrow due to the interference of machines. Clinical use of the PHANToM system is impossible at this moment, because it is difficult to feedback accurate haptic information to surgeons dealing with complicated forceps structures during endoscopic surgery. Consequently, novel approaches are necessary to introduce haptics into robotic surgery.

Table 1. Reports of force feedback systems

Force feedback system	Year / References
Piezoelectric system	2005 / 22
Electromagnetic system	2004 / 23
Servo motors	1994 / 24
Pneumatic haptic feedback actuator array	2007 / 25
PHANToM	2006 / 21, 26
Friction compensation system with linear motor	2005 / 36

Our Basic Study of Force Feedback

In routine open or endoscopic surgery, knot-tying force is decided intuitively based on the experience of the surgeon and the sense of touch imparted to the hands from the tissues being handling. In robotic surgery, however, this force is decided based on visual information alone, such as the stretch of the thread and tissue deformation. Wound healing at a site of anastomosis in the gastrointestinal tract and vessels greatly influences functional recovery of reconstructed organs postoperatively. We, therefore, advocate that objective data be obtained to determine the most suitable knot-tying force for appropriate suturing during robotic surgery. We speculated that estimating the most appropriate knot-tying force for tissues, in terms of the efficiency of wound healing, would make it possible to obtain basic in vivo data for the development of robotic forceps imparting a sense of touch to the surgeon's hands. It has been reported that wound healing in the gastrointestinal tract may be closely related to angiogenesis. [28,29], and that various growth factors may be involved in the process of wound healing. [30,31] Some studies have indicated that basic-fibroblast growth factor (bFGF) may take part in the repair of epithelial cells, especially in the stomach and intestine. [32,33] In this study, we examined angiogenesis and the expressions of growth factors in the area of ligation as parameters of wound healing, to evaluate the most appropriate knot-tying force for tissues during surgery. Furthermore, we determined the knot-tying force causing tissue ischemia by real-time measurement of local blood flow.

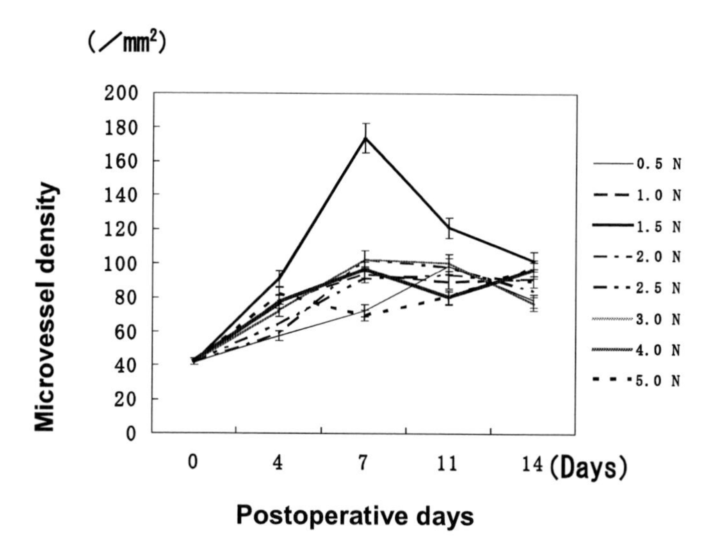

Figure 1. Relation between knot-tying force and the density of microvessels in the submucosa of the stomach. (ref. 34).

For knot-tying forces of 1.5 N or less, there was an inverse correlation between the knot-tying force and local blood flow. On the 7th and 11th PODs in the stomach and the jejunum, respectively, the density of microvessels in the submucosa at the sites of cutting and ligation was higher for the knot-tying force of 1.5 N than for any other force used (Figure 1). On the 4th and 7th PODs in the stomach and the 11th POD in the jejunum, the density of bFGF-positive cells in the mucosa at the sites of cutting and ligation was higher for the knot-tying force of 1.5 N than for any other force used (Figure 2). By assessing the efficiency of wound healing in terms of the extent of angiogenesis and growth factor expressions at wound sites, we determined that a knot-tying force of 1.5 N may be most appropriate for optimal wound healing in the gastrointestinal tract [34].

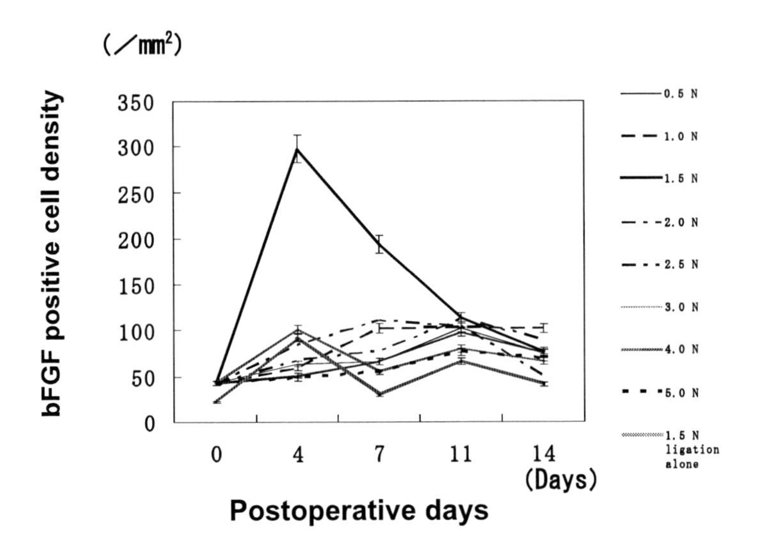

Figure 2. Relation between knot-tying force and the density of bFGF positive cells in the mucosa of the stomach. (ref. 34).

In recent years, the development and application of robotic surgery has progressed to such an extent that the poor flexibility of the tip of the forceps, which is one of the weak points of endoscopic surgery, has been overcome, making telesurgery possible, and suturing and ligation can now be performed more smoothly. [13,35] On the other hand, a sense of actually touching the tissues is not imparted to the surgeon's hands from the tip of the forceps, making use of the appropriate force for ligation difficult. We developed a system in which information about the force at the tip of the forceps is displayed to the surgeon on a monitor or via auditory signals. Our results suggest that suturing and ligation in the gastrointestinal tract using the appropriate force is possible using this system. We are now planning to develop a system in which the force at the tip of the forceps would be directly imparted, as a sense of touch, to the surgeon's hands. It is important that the sense of touch, which is different from other human senses, be bilateral. The sense of touch, which we are always aware of, consists of various types of information such as the nature, elasticity and so on at the surface of the substance being touched. Humans usually receive information processed as a return function and at the same time also add force application to the subject. We analyzed this information as combination of location and force. In other words, we analyzed it as information on location, speed and acceleration. The presence of the friction is known to greatly influence the precision and analysis of force feedback. If a high quality sensor is used, the friction in the system will prevent accurate force feedback. Friction compensation systems were thus developed, using a pipe-shaped linear actuator to drive the tip of the forceps, in the department of science and technology of our university (Figure 3) [36]. A bilateral teleoperation system was separated into common and differential modes using an acceleration-based controller. The servo force was attained in the common mode and the position error was regulated in the differential mode. In order to consider the conformity of force according to position, the servo force and the position regulator were integrated for acceleration. The acceleration-based controller was achieved by using a disturbance observer. The disturbance observer can detect external force as acceleration information. This forceps achieved a wide frequency response for high force reproducibility. This system can be used for the telesurgery because it is a master-slave type. As various elastic bodies were subjected to forces and consequent strains were detected, it became clear that the relation of both was $y=ae^{-bx}$. We would need to program the information of variable senses of touch into this system to create a haptic database. We believe that the results of our study provide useful data for surgery on living beings and will contribute to the future development of robotic forceps with a sense of touch. Our study is only an example showing the appropriate knot-tying force for wound healing of the gastrointestinal tract. In the future, surgeons should investigate the knot-tying and grasping forces ideally suited to various organs and tissues during surgery, and thereby provide useful in vivo data for surgery.

The Future of Robotic Surgery

Surgical robotics aims for the performance of minimally invasive surgery safely. Increasing complications with robotic surgery would be a sign of incorrect priorities. Thus, the development of haptics for robotic surgery requires urgent advancement. Unfortunately, there is presently no avenue for clinical use. It is necessary for surgeons and engineers to work together closely to develop haptics for robotic surgery. We await the realization of sensorless haptics in order to develop more precise haptic forceps, and we must obtain basic examination information about haptics. These technologies will be applicable to various fields, not only robotic surgery but also palpation, medical education and engineering science.

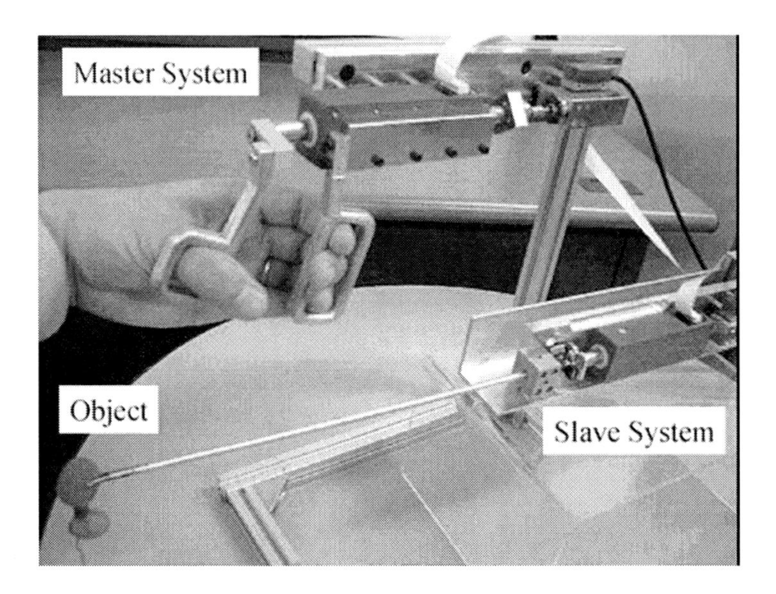

Figure 3. Haptic forceps system (ref. 37).

References

[1] Steiner CA, Bass EB, Talamini MA et al. Surgical rates and operative mortality for open and laparoscopic cholecystectomy in Maryland. *N Engl J Med* **330**:402-408, 1994.

[2] Escarce JJ. Externalities in hospitals and physician adoption of a new surgical technology: an exploratory analysis. *J Health Econ* **15**:715-734, 1996.

[3] Weeks JC, Nelson H, Gelber S et al. Short-term quality-of-life outcomes following laparoscopic-assisted colectomy vs open colectomy for colon cancer: A randomized trial. *JAMA* **287**: 321-328, 2002.

[4] Satava RM. Surgical robotics: the early chronicles: a personal historical perspective. *Surg Laparosc Endosc Percut Tech* 2002;12:6-16.

[5] Cleary K, Nguyen C. State of the art in surgical robotics: Clinical applications and technology challenges. *Comput Aided Surg* **6**: 312-328, 2001.

[6] Paul HA, Bargar WL, Mittlestadt B et al.. Development of a surgical robot for cementless total hip arthroplasty. *Clin Orthop* **285**: 57-66, 1992.

[7] Unger SW, Unger HM, Bass RT: AESOP robotic arm. *Surg Endosc* **8**: 1131, 1994.

[8] Schurr MO, Arezzo A, Neisius B et al.: Trocar and instrument positioning system TISKA. An assist device for endoscopic solo surgery. *Surg Endosc* **13**: 528-531, 1999.

[9] Buess GF, Arezzo A, Schurr MO et al. A new remote- controlled endoscope positioning system for endoscopic solo surgery. The FIPS endoarm. *Surg Endosc* **14**: 395-399, 2000.

[10] Aiono S, Gilbert JM, Soin B et al. Controlled trial of the introduction of a robotic camera assistant (EndoAssist) for laparoscopic cholecyctectomy. *Surg Endosc* **16**: 1267-1270, 2002.

[11] Schurr MO, Buess G, Neisius B et al. Robotics and telemanipulation technologies for endoscopic surgery. A review of the ARTEMIS project. Advanced Robotic Telemanipulator for Minimally Invasive Surgery. *Surg Endosc* **14**: 375-381, 2000.

[12] Falk V, Diegeler A, Walther T et al. Total endoscopic computer enhanced coronary artery bypass grafting. *Eur J Cardiothorac Surg* 17: 38-45, 2000.

[13] Cadiere GB, Himpes J, Germay O et al. Feasibility of robotic laparoscopic surgery: 146 cases. *World J Surg* **25**: 1467-1477, 2001.

[14] Reichenspurner H, Damiano RJ, Mack M et al. Use of the voice-controlled and computer-assisted surgical system ZEUS for endoscopic coronary artery bypass grafting. *J Thorac Cardiovasc Surg:* **118**: 11-16, 1999.

[15] Jourdan IC, Dutson E, Garcia A et al: Stereoscopic vision provides a significant advantage for precision robotic laparoscopy. *Br J Surg* **91**: 879-885, 2004.

[16] Ruurda JP, Wisselink W, Cuesta MA et al. Robot-assisted versus standard videoscopic aortic replacement: A comparative study in pigs. *Eur J Vasc Endovasc Surg* **27**: 501-506, 2004.

[17] Hashizume M, Shimada M, Tomikawa M et al. Early experiences of endoscopic procedures in general surgery assisted by a computer-enhanced surgical system. *Surg Endosc* **16**: 1187-1191, 2002.

[18] Sung GT, Gill IS: Robotic laparoscopic surgery: A comparison of the daVinci and ZEUS system. *Urology* **58**: 893-898, 2001.

[19] Bethea BT, Okamura AM, Kitagawa M et al. Application of haptic feedback to robotic surgery. *J Laparoendosc Adv Surg Tech A* **14**: 191-195, 2004.

[20] Okamura A, Smaby N, Cutkosky M. An overview of dexterous manipulation. *IEEE International Coference on Robotics and Automation* **1**: 255–262, 2000.

[21] Thorey G, Desai JP, Castellanos AE. Force feedback plays a significant role in minimally invasive surgery. *Ann Surg* **241**: 102-109, 2005.

[22] Omata S, Muratama Y, Constantinou CE: Multi-sonsory surgical support system incorporating, tactile, visual and auditory perception modalities. *Medicine Meets Visual Reality* **13**: 369-371, 2005.

[23] Hanson B, Levesley M: Self-sensing applications for electromagnetic actuators. *Sensors and Actuators* **116**: 345-351, 2004.

[24] Engel FL, Goossens P, Haalma R: Improved efficiency through I- and E-feedback: A trackball with contextual force feedback. *International Journal of Human-Computer Studies* **41**: 949-974, 1994.

[25] King CH, Higa AT, Culjat MO et al. A pneumatic haptic feedback actuator array for robotic surgery or simulation. *Medicine Meets Virtial Reality* **15**: 217-222, 2007.

[26] Tavakoli M, Aziminejad A, Patel RV et al. Methods and mechanisms for contact feedback in a robot –assisted minimally invasive environment. *Surg Endosc* **20**: 1570-1579, 2006.

[27] SenaAble Technologies, Inc. PHANToM: http://www.sensable.com//.

[28] Frank A, David CW: Angiogenesis in wound healing. *Pharmac Ther* 1991; 52: 407-422.

[29] Nicola JB, Edward AES, Simon SC, Malcolm RR: Aniogenesis induction and regression in uman surgical wound. *Wound Rep Reg* 2002; 10: 245-251.

[30] Glenn FP, John ET, Donna Y Thomas AM, Gary MF, Arlen T: Platelet-derived growth factor (BB homodimer), transforming growth factor-β1, and basic fibroblast growth factor in dermal wound healing. *Am J Pathol* 1992; 140: 1375-1388.

[31] Michael AB, Guido P, Scott G, Rolando HR: Temporal expression of TGF-β1, EGF, and PDGF-BB in a model of colonic wound healing. *J Surg Res* 1998; 80: 52-57.

[32] Dignass AU, Tsunekawa S, Podolsky DK: Fibroblast growth factors modulate intestinal epithelial cell growth and migration. *Gastroenterology* 1994; 106: 1254-1262.

[33] Terasaki T, Shimada K, Wakabayashi H, Tanaka M, Watanabe A: Study of the repairing of gastric ulcer using multivariate analysis of bFGF-positive cells, hemodynamics, PAS-positive mucus amount and glandular index in the gastric mucosa. *J Gastroenterol Hepatol* 1996; 11: 928-937.

[34] Oguma J, Ozawa S, Morikawa Y et al. Knot-tying force during suturing and wound healing in the gastrointestinal tract. *J Surg Res* **140**: 129-134, 2007.

[35] Ballantyne GH: Robotic surgery, telerobotic surgery, telepresence, and telementoring. Review of early clinical results. *Surg Endosc* **16**: 1389-1402, 2002.

[36] Katsura K, Iida W, Ohnishi K: Medical Mechatronics – An application to haptic forceps. *Annual Review in Control* **29**: 237-245, 2005.

In: New Robotics Research
Editors: E.D. Wagner et al, pp. 13-31
ISBN: 978-1-60741-093-5

Chapter 2

A HONEYBEE INSPIRED HETEROGENEOUS FORAGING SWARM MODEL

***Jia Song*[1,2,*], *Shengwei Yu*[3,†] *and Li Xu*[1,‡]**
[1]College of Electrical Engineering, Zhejiang University, P.R. China
[2]Department of Electronic Engineering, Suzhou Vocational College, P.R. China
[3]School of Electrical and Computer Engineering, Purdue University, USA

Abstract

Inspired by the organized behaviors of honeybee swarms, an individual-based mathematical model is proposed in this chapter for the heterogeneous swarm. The heterogeneous swarm is assumed to consist of two different kinds of individuals, namely, the scouts and the normal agents, with respect to their sensing abilities. Besides, a short-distance-bounded-attraction function was proposed to describe the attraction among individuals.

Firstly the heterogeneous swarm model is identified and the swarm cohesion is proved, and the analytical bound on the swarm size is provided. Secondly, the foraging properties of the heterogeneous swarm in multimodal Gaussian environment are studied, and conditions for collective convergence to more favorable regions are provided. Thirdly, simulations were carried out and the priority of proposed short-distance-bounded-attraction function was demonstrated in complex environment. Simulation results show that the

[*] E-mail address: sjia@jssvc.edu.cn
[†] E-mail address: shinewaysw@hotmail.com
[‡] E-mail address: xupower@zju.edu.cn

heterogeneous swarm model provides a feasible framework for multi-robot navigation applications.

Keywords: honeybee, heterogeneous swarm, foraging property, short-distance-repulsion

1. Introduction

Animals that travel in groups often rely on interactions among group members to make movement decisions. Social insects are particularly impressive examples, relying on interactions among nest mates they can maintain cohesive behaviors and appropriately respond to environment stimuli. These behaviors have certain advantages such as enhancing the chances of finding food and avoiding predators. In recent years, a variety of efforts have been devoted to modeling and analyzing the swarm behavior, hoping to gain similar advantages on control of multi-agent systems such as multiple mobile robots, autonomous flying machine, etc.

Stephens's paper [1] is one of the early works proposed the foraging theory in animal behavior. In [2], based on the foraging behavior of ant colony, an ant colony optimization algorithm was proposed. Passino[3] and his coworkers used a bacteria inspired model for a swarm moving in an environment with an attraction/repulsion profile. In [4], Gazi and Passino improved their earlier model[3] by adding artificial potential function to the inter-individual interactions and the interactions with the environment, the stability property of the swarm cohesion for different profiles were studied. This model can be viewed as a representation of cohesive social foraging swarms. Almost all the models proposed so far hypothesize that the swarms are homogeneous, in other words, the swarms are composed of the same type of agents with the same functions. For instance, the foraging model in[4] assume that all the individuals use the same dynamic function. Therefore, a disadvantage is that all the individuals need to know the exact relative positions of other individuals and have the ability of sensing the environment. This result in the fact that as the number of the swarm members grows, the computation needed by each agent also grow linearly. In engineering application, such as a multi-robot system based on these models, all the robots need to have excellent detection ability and the production cost will increase accordingly.

Inspired by the organized behaviors of honeybee swarms, an individual-based mathematical model is proposed in this chapter for a heterogeneous swarm. The heterogeneous swarm is assumed to consist of two different kinds of individuals, namely, the scouts and the normal agents, with respect to their sensing abilities. Cohesion and foraging properties of the swarm model are studied. In addition, a short-distance-bounded-attraction function is proposed, and simulation results show that this attraction function can significantly improve the foraging accuracy. This swarm model may provide a feasible framework for the implementation of a multi-robot system with heterogeneous sensing capabilities. For instance, a small subset of advanced robots with powerful sensors can guide a large number of simple robots to targets and warn them of potential dangers [14].

2. Biology Evidence

A colony of honey bees can achieve a high level of organization via dynamic division of labor and social interaction. Although there is no leader, the colony is very effective in foraging, comb construction, hive defense, thermoregulation, and other activities[5][11][12]. For example, an intriguing feature of the flight of a honey bee swarm is that only approximately 5% of the bees in swarm have visited the new nest prior to swarm lift off (Seeley et al. 1979). Nevertheless, in the majority of cases a swarm will fly quickly and directly to its destination. Two mechanisms of swarm guidance have been proposed [6]. Lindauer (1955) observed in airborne swarms that some bees fly through the swarm cloud with high speed, seemingly ‘pointing’ the direction to the new nest site. Lindauer suggested that these fast-flying bees are scouts that have visited the chosen nest site, and that their behavior guides the other, uninformed bees towards their new home. Normally Lindauer’s hypothesis is referred as the vision hypothesis. An alternative to the vision hypothesis is the olfaction hypothesis of Avitabile et al. (1975), who proposed that the scouts provide guidance by releasing assembly pheromone from their Nasanov glands on one side of the swarm cloud, thereby creating an odour gradient that can guide the other bees in the swarm. Both the vision and the olfaction hypotheses of swarm guidance seems reasonable, but none of them has been tested empirically until Beekman[7] studied the flights of both normal honeybee swarms and swarms in which each bee’s Nasanov gland was sealed shut. The test result proves that only the vision hypothesis is the actual mechanism of swarm guidance.

The clustering in honey bees and in-transit honey bee swarms are spectacular phenomena that have been studied experimentally by biologists during the past several decades. However, many aspects of these phenomena are still not well understood. Here we are only interested in the swarming behavior of the bees after they lift off and the cohesion of the in-transit swarm while moving to the new nest site. Moreover, the inspiration given by honey bee behavior for our research is that in a heterogeneous swarm a small fraction of informed scout agents can successfully guide all the other uninformed normal agents to the destination.

3. Heterogeneous Swarm Model

We consider a heterogeneous swarm including M individuals in an n-dimensional Euclidean space, and model all the individuals as points and ignore their dimensions. Assume that the M individuals can be divided into two different kinds according to their sensing abilities: N ($N<M$) scouts and (M-N) normal individuals. Thus we define the scout rate is $\eta = N / M$. The position of individual i is described by $x^i \in \mathbb{R}^n$. The interactions with environment in this model are based on artificial potential functions, a concept that has been used extensively for robot navigation and control[8][10]. Let $\sigma : \mathbb{R}^n \to \mathbb{R}$ represent the artificial potential function that model the environment containing obstacles to avoid and targets to move towards. Take the obstacle as a high potential region and the destination as a low potential region. It is assumed that all the swarm members move simultaneously and know the exact relative positions of all the other members, but only the scouts have the ability of detecting the environment. The equation of motion for each individual i can be described by:

$$\begin{cases} \dot{x}^i = -\nabla_{x^i}\sigma\left(x^i\right) + \sum\limits_{j=1, j\neq i}^{M} g\left(x^i - x^j\right), & i = 1, \ldots, N \\ \dot{x}^i = \sum\limits_{j=1, j\neq i}^{M} g\left(x^i - x^j\right), & i = N+1, \ldots, M \end{cases} \tag{1}$$

The term $-\nabla_{x^i}\sigma\left(x^i\right)$ represents that the scout agents can detect the environment and move towards low potential regions (analogous to destination). $g(\cdot)$ represents the function of mutual attraction and repulsion between the individuals. According to the study results on swarming behavior, the effect of function $g(\cdot)$ should be attractive for large distance and repulsive for short distance. In general, the term of attraction/repulsion function that we consider is

$$g(y) = -y\left[g_a(\|y\|) - g_r(\|y\|)\right] \tag{2}$$

Where $g_a : \mathbb{R}^+ \to \mathbb{R}^+$ represents the magnitude of the attraction term , whereas $g_r : \mathbb{R}^+ \to \mathbb{R}^+$ represents the magnitude of the repulsion term. $\|y\|$ is the Euclidean norm $\|y\| = \sqrt{y^T y}$. One issue to note here is that the attraction/repulsion function $g(\cdot)$ is odd.

Here we consider a kind of linear-attraction and bounded-repulsion function

$$\begin{cases} g_a(\|y\|) = a, & a > 0 \\ g_r(\|y\|)\|y\| \le b, & b > 0 \end{cases} \tag{3}$$

This function is consistent with the characteristic of the interaction among biological individuals and has been broadly used in the research[4] [9][13]. In this chapter we will study the characteristics of heterogeneous swarm using such a function. The attraction/repulsion function can take the form

$$g(y) = -y\left[a - b\exp\left(-\frac{\|y\|^2}{c}\right)\right] \tag{4}$$

Where a, b and c are all positive constants such that $b>a$。

4. Analysis of Swarm Cohesion

In this section, we will analyze the cohesiveness of the swarm and try to find bounds on the ultimate swarm size. To this end, we define the distance between individual i and the swarm center as

$$e^i = x^i - \overline{x} \tag{5}$$

Where $\overline{x} = \frac{1}{M}\sum_{i=1}^{M} x^i$. Therefore, the maximum distance from scouts to the swarm center can be described as

$$e_{s\max} = \max_{i=1,\ldots,N} \left\| e_i \right\| \tag{6}$$

Let the Lyapunov function for each individual be

$$V_i = \frac{1}{2}\left\| e^i \right\|^2 \tag{7}$$

Substituting equation (3) and taking the time derivation of V_i, we obtain

$$\dot{V}_i \leq$$

$$\begin{cases} -aM\left\|e^i\right\|^2 + \sum_{j=1,j\neq i}^{M} g_r\left(\left\|x^i - x^j\right\|\right)\left\|x^i - x^j\right\|\left\|e^i\right\| + \left\|\nabla_{x^i}\sigma\left(x^i\right) - \frac{1}{M}\sum_{j=1}^{N}\nabla_{x^j}\sigma\left(x^j\right)\right\|\left\|e^i\right\|, \\ (i = 1,\ldots,N) \\ -aM\left\|e^i\right\|^2 + \sum_{j=1,j\neq i}^{M} g_r\left(\left\|x^i - x^j\right\|\right)\left\|x^i - x^j\right\|\left\|e^i\right\| + \frac{1}{M}\left\|\sum_{j=1}^{N}\nabla_{x^j}\sigma^T\left(x^j\right)\right\|\left\|e^i\right\|, \\ (i = N+1,\ldots,M) \end{cases}$$

Note that the gradient of almost any realistic profile (e.g., plane and Gaussian profiles) is bounded. Thus, it is reasonable to have the following assumption about the profile.

Assumption 1: There exist a constant $\bar{\sigma} > 0$ such that $\left\|\nabla_y \sigma\left(y\right)\right\| \leq \bar{\sigma}, \forall y$

Then, for $i = 1,\ldots,N$

$$\begin{aligned} &\left\|\nabla_{x^i}\sigma\left(x^i\right) - \frac{1}{M}\sum_{j=1}^{N}\nabla_{x^j}\sigma\left(x^j\right)\right\| = \frac{1}{M}\left\|\left(M-1\right)\nabla_{x^i}\sigma\left(x^i\right) - \sum_{j=1,j\neq i}^{N}\nabla_{x^j}\sigma\left(x^j\right)\right\| \\ &\leq \frac{1}{M}\left[\left(M-1\right)\bar{\sigma} + \left(N-1\right)\bar{\sigma}\right] = \frac{M+N-2}{M}\bar{\sigma} \end{aligned} \tag{8}$$

Since $g_r\left(\left\|y\right\|\right)\left\|y\right\| \leq b$, we have

$$\dot{V}_i \leq \begin{cases} -aM\left\|e^i\right\|\left[\left\|e^i\right\| - \frac{b(M-1)}{aM} - \frac{M+N-2}{aM^2}\bar{\sigma}\right], i = 1,\dots,N \\ -aM\left\|e^i\right\|\left[\left\|e^i\right\| - \frac{b(M-1)}{aM} - \frac{N}{aM^2}\bar{\sigma}\right], i = N+1,\dots,M \end{cases} \tag{9}$$

Therefore, we conclude the following results:

Lemma 1: Using the linear-attraction and repulsion-bounded function given in (3), consider the heterogeneous swarm described by the model in (1) with scout rate $\eta = N / M$. Assume that the environment satisfies Assumption 1. Then, as $t \to \infty$ the scout's position $x^i(t) \to B_{\varepsilon_1}(\bar{x}(t))$, $i = 1,\dots,N$, whereas the normal agent's position $x^i(t) \to B_{\varepsilon_2}(\bar{x}(t)), i = N+1,\dots,M$. Where

$$B_{\varepsilon_k}(\bar{x}(t)) = \left\{y(t) : \left\|y(t) - \bar{x}(t)\right\| \leq \varepsilon_k\right\}, k = 1,2$$

$$\begin{aligned} \varepsilon_1 &= \frac{M-1}{aM}\left[b + \frac{M(1+\eta)-2}{M(M-1)}\bar{\sigma}\right] \\ \varepsilon_2 &= \frac{M-1}{aM}\left[b + \frac{\eta}{(M-1)}\bar{\sigma}\right] \end{aligned} \tag{10}$$

This result is important because it proves the cohesiveness of the heterogeneous swarm and provides an upper bound on the swarm size, which is defined as the radii of the hyperball centered at $\bar{x}(t)$ and containing all the individuals. According to the results above, the ultimate swarm sizes depend on the inter-individual attraction/repulsion parameters (*a* and *b*), the parameter of the environment ($\bar{\sigma}$), and etc. Note that the dependence on these parameters makes intuitive sense. Larger attraction (Larger *a*) leads to a smaller swarm size. In contrast, larger repulsion (larger *b*) or faster changing landscape (larger $\bar{\sigma}$) leads to a larger swarm size and these are intuitively expected results.

Note that the swarm size decreases with the increasing of individual number *M*. This is consistent with some biological swarms, where it has been observed that individuals are attracted to larger swarms. However, in biological swarms the number of the members *M* can be very large and as $M \to \infty$ both

ε_1 and ε_2 approach constant values. This implies that for large values of *M*, the size of the cohesive swarm is relatively independent of the number of the individuals, the scout rate and the characteristics of environment. Note also that while $M \geq 2$ we have $\varepsilon_1 \geq \varepsilon_2$. This implies that, compared with normal individuals, the scouts are farer away from the swarm center. This is because that the scouts can detect the environment, whereas normal agents have limited sensing ability and can only get information about the environment indirectly from the scouts. It means that the link between the environment and scouts is more direct and larger. Therefore, it is much easier for the scouts to be drawn away from the swarm center than the normal agents.

Note also that the scout rate η can only take effect together with environment parameter $\bar{\sigma}$, and larger η (more scouts) leads to a larger swarm size. This is because with larger number of scouts the swarm can obtain more information about environment and thereby the swarm size will be increased. However, decreasing the scout rate η will not have negative effect on swarm cohesiveness. On the contrary it may decrease the swarm size. This result is what we expect and proves the feasibility and rationality of the proposed heterogeneous swarm model.

5. Analysis of Foraging Behavior

In this section we consider the foraging behavior of heterogeneous swarm in realistic profile. The profile we used is a combination of Gaussian profiles. In other words, we consider the profile given by

$$\sigma(y) = -\sum_{i=1}^{K} \frac{A_\sigma^i}{2} \exp\left(- \frac{\left\| y - c_\sigma^i \right\|^2}{l_\sigma^i} \right) + b_\sigma \tag{11}$$

Where $A_\sigma^i \in \mathbb{R}, b_\sigma \in \mathbb{R}, l_\sigma^i \in \mathbb{R}^+, c_\sigma^i \in \mathbb{R}^n$ for all $i = 1, \ldots, K$. Note that since the A_σ^i can be positive or negative, there can be both hills and valleys leading to a complex environment.

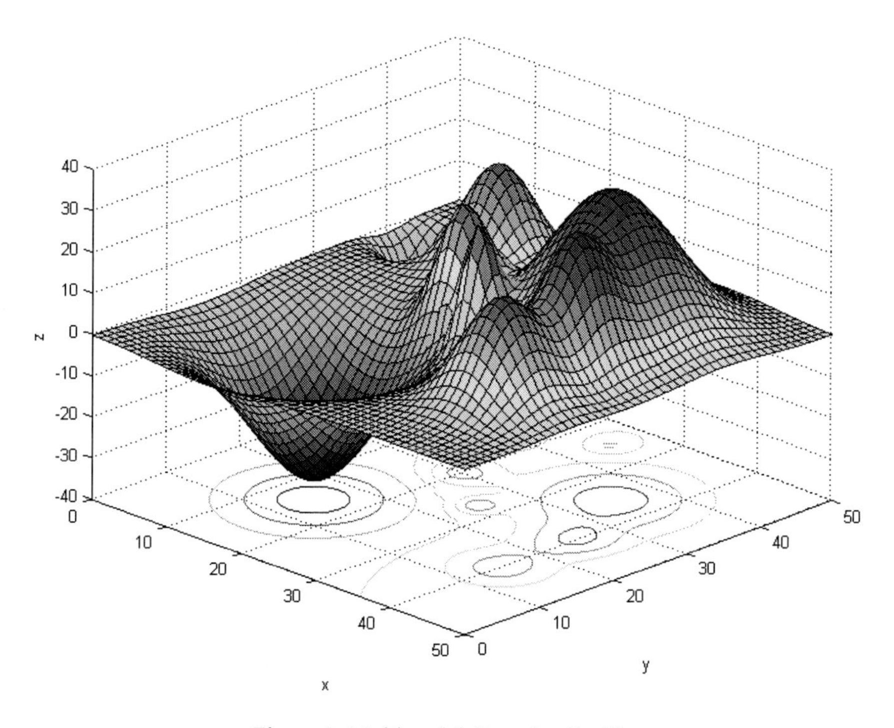

Figure 1. Multimodal Gaussian Profile.

Note that for this profile Assumption 1 is satisfied with

$$\bar{\sigma} = \sum_{i=1}^{K} \frac{\left|A_{\sigma}^{i}\right|}{\sqrt{2l_{\sigma}^{i}}} \exp\left(-\frac{1}{2}\right) \tag{12}$$

Therefore Lemma 1 holds and the swarm cohesion in this profile is proved. Moreover, as $t \to \infty$, we will have

$$\begin{aligned} \varepsilon_{mG1} &= \frac{M-1}{aM}\left[b + \frac{M(1+\eta)-2}{M(M-1)} \sum_{i=1}^{K} \frac{\left|A_{\sigma}^{i}\right|}{\sqrt{2l_{\sigma}^{i}}} \exp\left(-\frac{1}{2}\right)\right] \\ \varepsilon_{mG2} &= \frac{M-1}{aM}\left[b + \frac{\eta}{(M-1)} \sum_{i=1}^{K} \frac{\left|A_{\sigma}^{i}\right|}{\sqrt{2l_{\sigma}^{i}}} \exp\left(-\frac{1}{2}\right)\right] \end{aligned} \tag{13}$$

This result specifies the regions in which the scouts and normal agents will converge respectively. In order to study the qualitative properties of swarm foraging behavior, we need to consider the motion of the swarm center. Firstly the distance from the swarm center $\overline{x}$ to the destination point c_σ is defined,

$$e_\sigma = \overline{x} - c_\sigma \tag{14}$$

Using Lyapunov function

$$V_\sigma = \frac{1}{2}\left\|e_\sigma\right\|^2 = \frac{1}{2}e_\sigma^T e_\sigma \tag{15}$$

Then, calculate the time derivative of it, in the case of $A_\sigma^k > 0$ we obtain

$$\dot{V}_\sigma^k \le -\left[\frac{A_\sigma^k}{Ml_\sigma^k}\sum_{i=1}^{N}\exp\left(-\frac{\left\|x^i - c_\sigma^k\right\|^2}{l_\sigma^k}\right)\right]\left\|e_\sigma^k\right\| \times \left[\left\|e_\sigma^k\right\| - e_{s\max} - \frac{\sum_{j=1,j\neq k}^{K}\frac{\left|A_\sigma^j\right|}{Ml_\sigma^j}\sum_{i=1}^{N}\exp\left(-\frac{\left\|x^i - c_\sigma^j\right\|^2}{l_\sigma^j}\right)\left\|x^i - c_\sigma^j\right\|}{\frac{A_\sigma^k}{Ml_\sigma^k}\sum_{i=1}^{N}\exp\left(-\frac{\left\|x^i - c_\sigma^k\right\|^2}{l_\sigma^k}\right)}\right]$$

Suppose that all the scouts are near c_σ^k and far from other minima. Then we have the following assumption.

Assumption 2: At one minimum c_σ^k, the positions of scouts satisfy the following conditions

$$\begin{cases} \left\|x^i - c_\sigma^k\right\| \le h_k\sqrt{l_\sigma^k},\ h_k > 0, \\ \left\|x^i - c_\sigma^j\right\| \ge h_j\sqrt{l_\sigma^j},\ h_j \ge \sqrt{\frac{1}{2}},\ j = 1,\dots,K,\ j \neq k\ \forall i = 1,\dots,N \\ \frac{A_\sigma^k}{\sqrt{l_\sigma^k}}h_k\exp\left(-h_k^2\right) > \sum_{j=1,j\neq k}^{K}\frac{\left|A_\sigma^j\right|}{\sqrt{l_\sigma^j}}h_j\exp\left(-h_j^2\right) \end{cases} \tag{16}$$

Here h_k, h_j are some positive constants and satisfy the above condition. Therefore

$$\frac{\sum_{j=1,j\neq k}^{K}\frac{\left|A_{\sigma}^{j}\right|}{Ml_{\sigma}^{j}}\sum_{i=1}^{N}\exp\left(-\frac{\left\|x^{i}-c_{\sigma}^{j}\right\|^{2}}{l_{\sigma}^{j}}\right)\left\|x^{i}-c_{\sigma}^{j}\right\|}{\frac{A_{\sigma}^{k}}{Ml_{\sigma}^{k}}\sum_{i=1}^{N}\exp\left(-\frac{\left\|x^{i}-c_{\sigma}^{k}\right\|^{2}}{l_{\sigma}^{k}}\right)}\leq\frac{\sum_{j=1,j\neq k}^{K}\frac{\left|A_{\sigma}^{j}\right|}{\sqrt{l_{\sigma}^{j}}}h_{j}\exp\left(-h_{j}^{2}\right)}{\frac{A_{\sigma}^{k}}{l_{\sigma}^{k}}\exp\left(-h_{k}^{2}\right)}$$

Which implies that if $\left\|e_{\sigma}^{k}\right\|>e_{s\max}+\frac{\sum_{j=1,j\neq k}^{K}\frac{\left|A_{\sigma}^{j}\right|}{\sqrt{l_{\sigma}^{j}}}h_{j}\exp\left(-h_{j}^{2}\right)}{\frac{A_{\sigma}^{k}}{l_{\sigma}^{k}}\exp\left(-h_{k}^{2}\right)}\triangleq e_{s\max}+\varepsilon_{\sigma}$,

we can obtain $\dot{V}_{\sigma}^{k}<0$. Therefore, the swarm center will move toward c_{σ}^{k}.

Then, consider the heterogeneous swarm with condition in (16) satisfied. As $t\to\infty$, we have $\left\|e_{\sigma}^{k}\right\|\leq e_{s\max}+\varepsilon_{\sigma}$. Since $e_{s\max}\leq\varepsilon_{mG1}$, combine with the bound implied by (13), we obtain

$$\begin{aligned}\left\|x^{i}-c_{\sigma}^{k}\right\|&\leq\left\|e^{i}\right\|+\left\|e_{\sigma}^{k}\right\|\\&\leq\begin{cases}2\varepsilon_{mG1}+\varepsilon_{\sigma}, & i=1,\ldots,N\\ \varepsilon_{mG1}+\varepsilon_{mG2}+\varepsilon_{\sigma}, & i=N+1,\ldots M\end{cases}\end{aligned}\tag{17}$$

Thereby, we have the following lemma.

Lemma 2: Consider the heterogeneous swarm described in (1) with the inter-individual attraction/repulsion function as given in (3). Assume that the profile of environment is defined by (11). Moreover, suppose that the locations of scouts satisfy Assumption 2, as $t\to\infty$, we have:

In the case $A_{\sigma}^{k}>0$, for the scouts we have $x^{i}(t)\to B_{2\varepsilon_{mG1}+\varepsilon_{\sigma}}\left(c_{\sigma}^{k}\right), i=1,\ldots,N$, for the normal agents we have $x^{i}(t)\to B_{\varepsilon_{mG1}+\varepsilon_{mG2}+\varepsilon_{\sigma}}\left(c_{\sigma}^{k}\right), i=N+1,\ldots,M$. Where

$$\varepsilon_{mG1} = \frac{M-1}{aM}\left[b + \frac{M(1+\eta)-2}{M(M-1)}\sum_{i=1}^{K}\frac{\left|A_\sigma^i\right|}{\sqrt{2l_\sigma^i}}\exp\left(-\frac{1}{2}\right)\right]$$

$$\varepsilon_{mG2} = \frac{M-1}{aM}\left[b + \frac{\eta}{(M-1)}\sum_{i=1}^{K}\frac{\left|A_\sigma^i\right|}{\sqrt{2l_\sigma^i}}\exp\left(-\frac{1}{2}\right)\right]$$

$$\varepsilon_\sigma = \frac{\sum_{j=1,j\neq k}^{K}\frac{\left|A_\sigma^j\right|}{\sqrt{l_\sigma^j}}h_j\exp\left(-h_j^2\right)}{\frac{A_\sigma^k}{l_\sigma^k}\exp\left(-h_k^2\right)}$$

This result implies that in multimodal Gaussian profile if all the scouts are initially close to one food resource and far from other resources, then all individuals of the heterogeneous swarm will converge to it eventually. Note that the accuracy of foraging depends on swarm cohesiveness and environment condition, but have no direct relation with the scout rate. In other words, the better cohesiveness leads to better foraging accuracy. Thus, with only a small fraction of scouts the swarm may still have good cohesiveness, and acquire high foraging accuracy. This again confirmed the rationality and feasibility of heterogeneous swarm using small number of scouts.

6. Short-Distance-Bounded-Attraction Function

Note that the attraction in (3) has no upper bound and increases linearly with the increasing of y. However, in biological swarms, it is common that the individuals' senses are limited and each individual can sense only the individuals in a limited range. Therefore, the attraction among individuals has an upper bound and will decrease when the distance is too large. In this section, we define a new attraction function called the short-distance-bounded-attraction function to overcome this problem. For the repulsion function we use the same type of function as in the previous section, i.e. functions satisfying (3). The interaction function that we consider is

$$g(y) = -y\left[a\exp\left(-\frac{\|y\|^2}{d}\right) - b\exp\left(-\frac{\|y\|^2}{c}\right)\right] \tag{18}$$

Where *a, b, c, d* are all positive constants such that $b>a$, $d>c$. Note that *c* and *d* represent the range of repulsion and attraction, respectively. The maximum of Attraction $g_a(\|y\|)\|y\|$ occurs at $\|y\| = \sqrt{\frac{d}{2}}$, and along with the increasing of distance $\|y\|$ the attraction will decrease until near 0. Therefore the attraction represented by this function has an upper bound, and is called short-distance-bounded-attraction. The upper bound and the range of attraction depend on parameter *a, d*. As compared to the linear-attraction function, our short-distance-bounded-attraction function is better consistent with the properties of the interaction among biology individuals, thus is a more reasonable and effective interaction function.

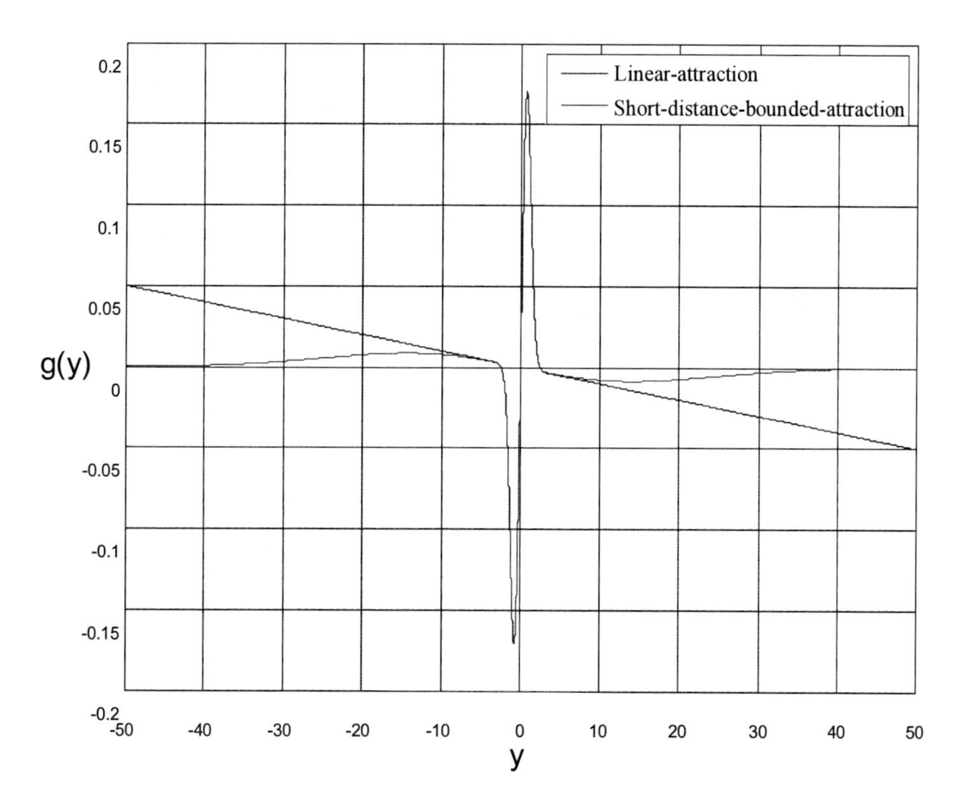

Figure 2. The comparison between short-distance-bounded-attraction function graphic and linear-attraction/bounded-repulsion function graphic.

The Figure 2 shows the plot of a short-distance-bounded-attraction/bounded-repulsion function (a=0.01, b=0.4, c=1.0, d=400) and the plot of a linear-attraction/bounded-repulsion function (a=0.01, b=0.4, c=1.0). As one can see, in short distance (small value of y) the two function plots look quite similar. Note that the attraction of short-distance-bounded-attraction/bounded-repulsion function will decrease while the inter-member distance is larger than 14, and the domain of this attraction is approximately less than 40.

It is easy to see that the short-distance-bounded-attraction and bounded-repulsion functions described in (18) satisfy Assumption 1. However, both the Lemma 1 and Lemma 2 need the attraction function satisfy $g_a(\|y\|) = a$, (a>0), and for the short-distance-bounded-attraction function, we can only find an a' in a certain range that satisfy $g_a(\|y\|) \geq a'$. Thus the two lemmas both hold only in this certain range and global convergence cannot be guaranteed. Later we will see that just because of this characteristic, the short-distance-bounded-attraction function can show better result than linear attraction function.

7. Simulation Results

In this section, we will provide some simulation examples to illustrate the theory developed in the preceding sections. The multimodal Gaussian profile we used is shown in Figure 1, which has three minima and six maxima. The global minimum is located at [15, 15], other two local minima are located at [5, 40] and [25, 45] respectively. In all the simulations performed below we chose the swarm with M=20 individuals and N=5 scouts, which means that the scout rate $\eta = 25\%$. As parameters of the linear-attraction and bounded-repulsion function in (2) we choose a=0.01, b=0.4, c=1.0.

The simulation results shown in Figure 3 are for the swarm using linear-attraction and bounded-repulsion function with scouts initially collecting around different positions in the environment. The normal individuals and the scouts are respectively represented by circles and triangles. The upper two plots in Figure 3 show two example runs for which we initialized all the scouts' positions nearby one minimum and far from other minima. For both of the simulations we can see that the entire swarm succeeds to converge to the minimum. Note that for these cases the Assumption 2 in the Lemma 2 is satisfied and the simulation results support the analysis of preceding sections. In addition, we observe that the center of scouts is overlapped with the minimum.

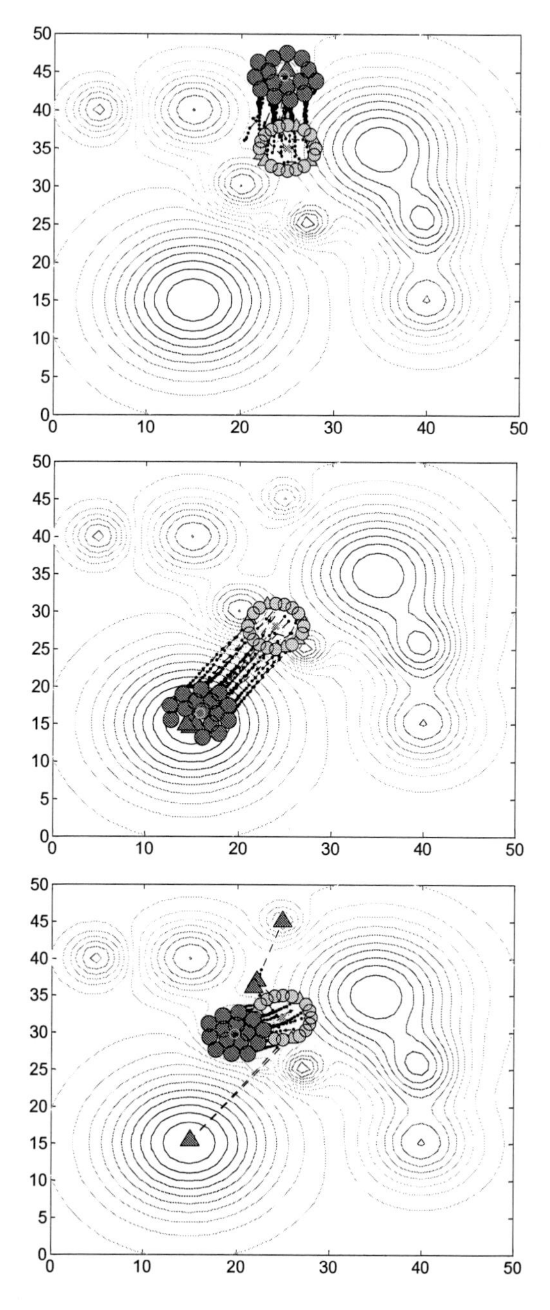

Figure 3. Simulation Results for swarm using linear-attraction and bounded-repulsion function (initial positions of scouts close to a minimum and the counter example).

If Assumption 2 is not satisfied, for instance, some scouts' positions are nearby one minimum while other scouts' positions are nearby another minimum, then the scouts will disperse (As being shown in the lower plot in Figure 3). Note that the normal individuals converge and stay at a position among the scouts (the attraction and repulsion from scouts balance), instead of moving towards any minimum.

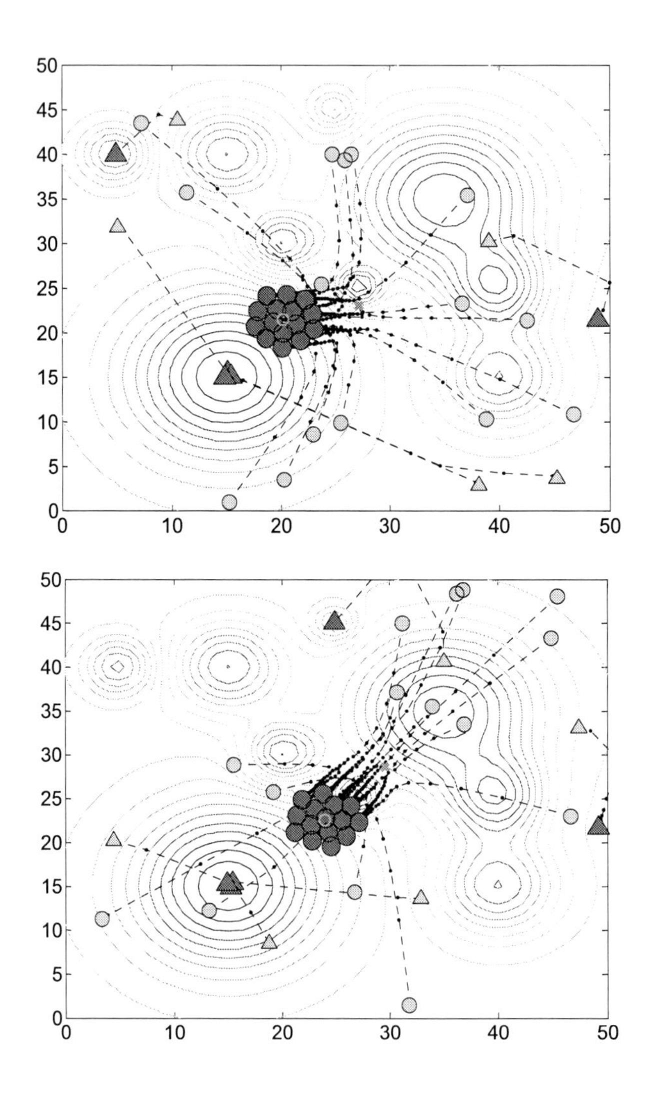

Figure 4. Simulation Results for swarm using linear-attraction and bounded-repulsion function (initial positions of individuals are random distributed).

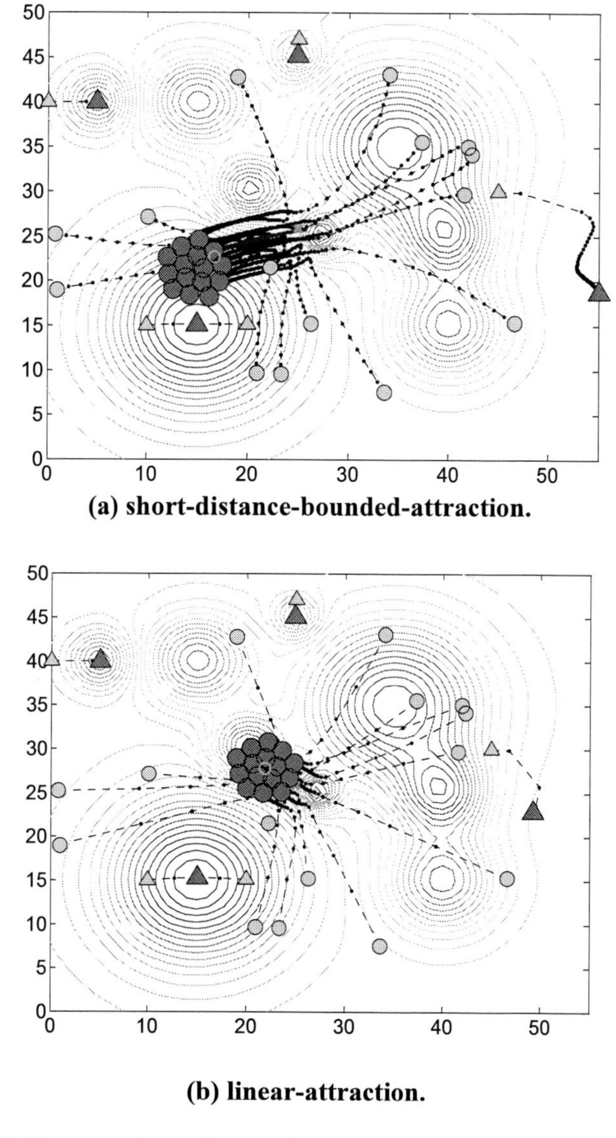

(a) short-distance-bounded-attraction.

(b) linear-attraction.

Figure 5. The comparison of swarm foraging result using different functions.

Figure 4 show the simulation results with initial positions chosen at random. Note that in this environment the region of the global minimum is the biggest. Therefore, if randomly distribute the swarm individuals in the profile, the probability that in the end the majority of scouts converge at global minimum will be the largest. As being shown in Figure 4, for both of the simulations we can see that three scouts (60% scouts) find the global minimum at last. Therefore for the normal individuals the attraction in the direction of the global minimum is strongest and this results in the fact that the entire swarm will move toward the global minimum.

Therefore, this result shows the advantage of the heterogeneous swarm partly including scouts. In this environment if all the swarm individuals are scouts, a certain percentage of individuals will move to regions far from the global minimum. If the number of individuals in the swarm is large, it will be a big loss. However, in contrast, a small fraction of scouts can successfully direct all the other normal individuals to the destination.

Now consider the interaction function described in (18), as parameters we choose a=0.01, b=0.4, c=1.0, d=400, then the maximum attraction 0.009 occurs at $\|y\| = 14.142$. In the preceding multi-modal Gaussian environment we place two scouts at the global minimum (it means that only 40% scouts can find the accurate food resource position), two scouts at two local minima respectively and the last scout at any other position. The initial positions of normal individuals are chosen at random. Comparing the simulation result of swarm using short-distance-bounded-attraction function (Figure 5 (a)) with the result using linear-attraction function (Figure 5(b)), we observe that the swarm center is much nearer to the global minimum in the first than the second case. Therefore, we can conclude that the simulation result of using short-distance-bounded-attraction function is much better than using linear-attraction function.

8. Conclusions

Inspired by the organized behaviors of honeybee swarms, an individual-based mathematical model is proposed in this chapter for the heterogeneous swarm. Cohesion and foraging properties of the heterogeneous swarm employs linear-attraction and bounded-repulsion functions in multimodal-Gaussian environment were proved. Furthermore, the priority of the proposed short-distance-bounded-attraction function was demonstrated. The simulation results prove that the heterogeneous swarm can eventually form an aggregation of finite size around swarm center, and converge to advantaged regions of the environment under certain conditions. The results show that a large number of uninformed normal agents can be guided successfully by only a small fraction of informed scouts. Our results might be particularly useful for designing and controlling multi-robot systems and mobile sensor networks, etc.

References

[1] D. W. Stephens and J. R. Krebs, *Foraging Theory.* Princeton, NJ: Princeton Univ. Press, 1986.

[2] E. Bonabeau, M. Dorigo, and G. Theraulaz, *Swarm Intelligence: From Natural to Artificial Systems.* New York: Oxford Univ. Press, 1999.

[3] K. M. Passino, Biomimicry of bacterial foraging for distributed optimization and control. *IEEE Control Syst. Mag.* vol. 22, pp. 52–67, June 2002.9.

[4] V Gazi, KM Passino, "Stability analysis of social foraging swarms," Systems, Man and Cybernetics, Part B, *IEEE Transactions on*, 2004, pp. 539–557.

[5] T. D. Seeley. *The Wisdom of the Hive: The Social Physiolog Colonies.* Harward University Press, Cambridge, Mass, 1995.

[6] S Janson, M Middendorf, M Beekman, "Honeybee swarms: how do scouts guide a swarm of uninformed bees?" *Animal Behaviour,* 2005, 70, 349–358.

[7] M Beekman, RL Fathke, TD Seeley, "How does an informed minority of scouts guide a honey bee swarm as it flies to its new home?" *Animal Behaviour,* 2006, 71, 161–171.

[8] J. H. Reif and H. Wang, "Social potential fields: a distributed behavioral control for autonomous robots," *Robot. Auton. Syst.,* vol. 27, pp. 171–194, 1999.

[9] Liang Chen, Li Xu, "Collective Behavior of an Anisotropic Swarm Model Based on Unbounded Repulsion in Social Potential Fields," *8th European Conference,* ECAL 2005, Canterbury, UK, September 5-9, 2005.

[10] Rimon, E., koditschek, D.E.: Exact Robot Navigation Using Artificial Potential Functions. *IEEE Trans. Robot. Automat.* **8**(1992) 501-518.

[11] T. D. Seeley and S. C. Buhrman. Group decision making in swarms of honey bees. *Behavioral Ecology and Sociobiology,* **45**:19{31, 1999.

[12] T. D. Seeley, R. A. Morse, and P. K. Visscher. The natural history of the flight of honey bee swarms. *Psyche,* **86**(2-3): 103{113, June-September 1979.

[13] V. Gazi and K. M. Passino, "Stability analysis of swarms," *IEEE Trans. Automat. Contr.*, vol. 48, pp. 692–697, Apr. 2003.

[14] V Kumar, D Rus, S Singh, "Robot and sensor networks for first responders," *Pervasive Computing, IEEE,* vol. 3, issue 4, pp.24—33, 2004.

In: New Robotics Research
Editors: E.D. Wagner et al, pp. 33-62

ISBN: 978-1-60741-093-5

Chapter 3

A PLATFORM FOR DEVELOPING AUTONOMY TECHNOLOGIES FOR SMALL MILITARY ROBOTS*

Gary Haas, Jason Owens and Jim Spangler

Abstract

In order to study autonomous behaviors in small military robots, researchers at the U.S. Army Research Laboratory (ARL) renovated an existing but outdated ATRV research robot. Commercial sensors with capabilities resembling those anticipated from the military tech base were selected and integrated, and the computing capability was substantially enhanced by judiciously selecting commercial components. Support electronics were upgraded or replaced as necessary. Safety elements common in larger robotic vehicles were integrated into the small ATRV chassis. Systems software was selected to provide a stable foundation for the advanced functions envisioned. Player, a middleware widely used by academic robot researchers, was incorporated as a springboard to the agent-based behaviors believed necessary for the next phase of development in robotics. A distributed development environment was implemented to enable parallel software efforts. Issues in software architecture were identified, and architectures from the literature were investigated in search of a foundation for future work. Without major investment, the antiquated research robot has become a key element in ARL's quest to develop technologies for a highly capable robot to team with soldiers on tomorrow's urban battlefield.

* Excerpted from Army Research Laboratory Report, ARL-MR-709, Updated December 2008

1. Objective and Program Background[1]

In late 2007, the U.S. Army Research Laboratory (ARL) initiated a research thrust targeting robotics technologies for asymmetric warfare in urban terrain. The vision for this tactical domain calls for small robots suited to maneuver indoors as well as outdoors, perhaps sharing space with troops, with sensors to perceive the immediate surroundings, and sufficient intelligence to enable (at least short periods of) unsupervised operation. Such a robot must embody substantial autonomy, e.g., be able to "see" and "understand" its environment so that it can perform its function with minimum burden on the soldiers it supports. It requires sensors capable of detecting the immediate surroundings with high fidelity and richness, and powerful onboard computing systems. At the inception of the new thrust, such capabilities were unavailable in research robots currently in ARL's robotics labs. As a first step toward exploring this new mission space, scientists and engineers at the Vehicle Technology Directorate's Unmanned Vehicle Technologies Division (UVTD) set out to create the capability.

A test bed for developing autonomy technologies, at least early in the program, can be based on commercial sensing technologies and a mobility platform of limited performance. The lab had in its inventory 8-year-old ATRV Jr. research robots once built by Real World Interface, Inc. (RWII). At the time of acquisition, these robots were quite advanced and offered skid-steer wheeled mobility, global positioning system (GPS) and electronic compass for navigation, ladar (a portmanteau of laser radar and often used interchangeably with lidar) and sonar sensors for obstacle detection, and a software development environment based on linked server modules. These robots were at the core of in-house robotics research at ARL. By 2007, the ATRVs were well worn. RWII (renamed iRobot Corporation) had discontinued production and support, and the Pentium III processor and Red Hat 6.2 operating system at the core of the robotics had been superceded by several generations.

Research supporting ARL's new robotics thrust calls for research robots similar in scale to the old ATRVs and with power and payload to support quantities of sensors and computing. New research robots have become available from several vendors, but, in general, the function of these products is not substantially different from that of the old ATRVs. The decision was made to

[1] The products described in this report are believed to be suitable for the intended use, but their use in this endeavor does not constitute an endorsement by the government. Other products may perform as well or better, or be less expensive.

renovate the old robots rather than invest in new research robots. This report describes the upgrade of the ATRV robot for its new role.

2. System Description

The stock ATRV robot is a 25-in-long x 24.5-in-wide x 21-in-high 110-lb vehicle. Two deep- draw, gel-cell, lead-acid batteries power a pair of servo motors that drive its four 12-in tires in skid-steer fashion by means of toothed belts. A sturdy rectangular sheet-metal chassis houses the internals and supports the mounting rails front and rear and on the deck.

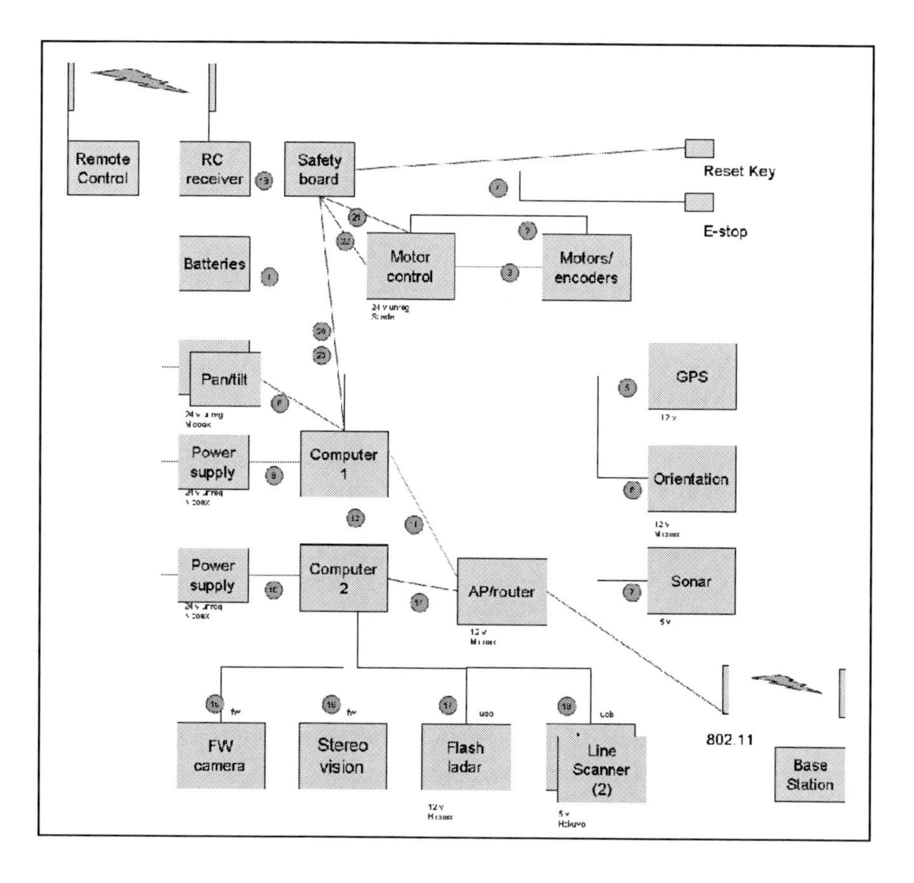

Figure 1. Major components of the upgraded ATRV. Gray indicates the component is off-board, linked by radio.

Of the rest of the original major components, only the pan-tilt unit remains. The rest have been replaced by functional equivalents and supplemented with functional extrapolations. Figure 1 depicts the components of the upgrade at a block diagram level. Details of the upgrade follow.

3. Upgrade

3.1. Hardware

3.1.1. Computational Hardware

Increasing the computational capacity was a requirement for the ATRV upgrades. In order to accommodate the planned sensor-processing research, the computers needed to represent the latest technology available off the shelf. In addition, they needed to be small and power-efficient but possess sufficient processing capability to minimize the processing bottleneck. The current trend in the hardware and software community is to leverage horizontal scale (more processing elements on a die), more so than clock speed. Thus the new computational hardware design takes full advantage of the space and processors available.

3.1.1.1. Central Processing Units (CPUs)

A Mini-ITX (a motherboard format popularized by Via Technologies, Inc. [*1*]) was located that supported dual- and quad-core Intel processors. The only one available at the time was the Commell Core 2 Quad[2] with dual gigabit ethernet ports, six USB 2.0 ports, an 8-bit digital general purpose input/output[3] port, and Serial Advanced Technology Attachment[4]. The ATRV Jr. has enough room for two reasonably sized mini-ITX cases, so the computers are relatively modular and easily replaceable. Each computer is equipped with 4 GB of RAM (although the computer system bus allows access to only 3 GB). A gigabit ethernet switch connects the two machines together and essentially yields a miniature computing cluster with eight processor cores.

[2] Model No. LV-676.
[3] Usually an 8-bit digital interface.
[4] A computer bus for attaching mass storage devices to a computer.

3.1.1.2. Storage

Storage is provided by a single 16-GB solid-state drive for each CPU. Solid-state drives are used to increase system performance (relative to standard rotating platter 4200 RPM laptop drives) and increase system reliability with respect to vibration issues. While the storage size is small compared to today's large drives, it is more than enough to hold an hour's worth of raw data from the primary sensors and can be easily expanded with larger sizes if the need arises.

3.1.1.3. Network

As mentioned previously, the computers interface with each other through the network switch and with external machines (i.e., control stations, logging/debugging systems, development systems) through either a 1 00-Mbs wired or 54-Mbs wireless interface provided by an Alfa Network's AWAP608 wireless access point [*2*].

3.1.1.4. Sensor Interfaces

The ATRV Jr. has a variety of onboard sensors that utilize several connection interfaces on the CPU: serial, USB 2.0, and IEEE 1394.[5] Many of the serial devices are connected through RS-232[6]-to-USB adapters, while others (like the motor controller) are connected directly to the motherboard. Several perception sensors on the ATRV Jr. use USB and IEEE 1394 directly.

3.1.2. Sensors for Reaction and World Modeling

The purpose of sensors on a mobile robot is to create an analog of the nearby environment in data structures used by the computer programs that control the robot. These data structures are collectively termed the "world model." While some robots simply react to sensor inputs to alter some behavior ("obstacle ahead, turn left"), the objective of this project is to enable the robot to sense the geometry of its environs, store the sensed elements in data structures based on a self- constructed local map, and plan its behaviors based on the map. The map will have different layers, populated with geometric elements (extracted and abstracted from its geometry sensors), spectral elements (extracted from its cameras), elements fused from the sensor-derived elements, and

[5] Also known as Firewire (another serial bus standard for computer interfaces).

iconic elements (extrapolated from the sensor-derived elements and filtered based on a mission-based context).

The universe of sensors appropriate for small robots is not a large one. Sensors considered for the robot are described further in this section, and those selected for the upgrade are pictured in figure 2.

3.1.2.1. Video Camera

Video cameras are widely available and relatively inexpensive. Imagery is dense (high pixel count) but only spectral in nature. The information content of the image is rich but lacks the immediate geometric significance needed for safe mobility. Significant processing is necessary to convert a stream of spectral images to the geometric world model needed. However, given a geometry by some other sensor, the richness of the video imagery can be overlaid. Video imagery is also the most easily interpreted sensor mode for a human. Augmented by feature-tracking software, the video sensor can provide a direction reference and can support algorithms such as direction-only simultaneous localization and mapping.

Given the widespread availability of video cameras, there are a number of parameters that can be used as a selection criterion. Field of view, a function of lens selection, is probably the most important (the wider the better). In general, the image resolution (number of pixels) is not important, as even the least capable cameras have sufficient resolution for the application (except for stereo vision, which will be treated separately). A key parameter is the ease of integration with a computer. While analog cameras today dominate the market, cameras with a built-in digital interface are more suitable for the application. This is partly due to ease of interfacing, but the primary reason is that most digital video cameras have progressive scan technology, which is important when the camera is mounted on a moving platform. Firewire (IEEE 1394) cameras are the most common. For streaming applications, Universal Serial Bus (USB) has little to offer over Firewire, which was designed with video applications in mind. Both offer 400-Mbs rates, adequate for video graphics array[7] (VGA)-quality resolution or a little more. The newer IEEE 1394b, at 800 Mbs, is not yet widely available but will become so.

[6] A standard serial interface to a computer, once common but recently supplanted by USB.
[7] A standard for computer display hardware.

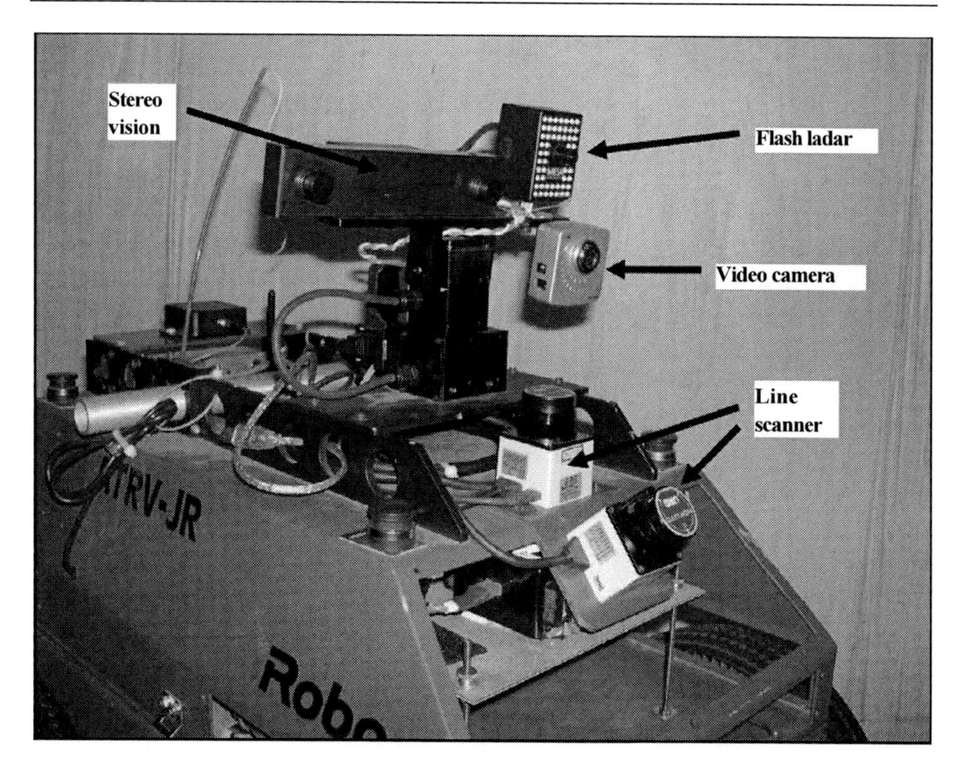

Figure 2. Principal sensor of the ATRV upgrade.

A representative video camera is the Unibrain Fire-i, priced at around $120 for VGA-resolution imagery with a plastic case and glass lens [*3*]. A 4.3-mm focal length *f* 2.0 lens was selected, specified to deliver VGA-resolution imagery at 30 Hz, covering a horizontal field of view of 42.25°. A number of vendors offer competing products; this selection was based on low price for adequate performance.

3.1.2.2. Near-Range Geometry Sensing

For a number of years the only sensor capable of detecting the geometry of the environment, in a package of suitable size and cost for use on a small robot, was a sonar sensor such as the Polaroid product of the early 1980s. Modern sonar sensors for robotics differentiate between empty and occupied volumes in a cone subtending as little as 15°, so the spatial resolution perpendicular to the cone is limited, and the "occupied" region of the cone can be a tiny fraction of the cross section of the cone at the detected range [*4*]. Sonar units work well enough for simple reactive obstacle detection, and several will be incorporated in the finished upgrade. However, a denser scan is necessary to be useful as a geometry sensor.

3.1.2.2.1. Line Scanner

A line scanner is one such sensor and is widely used in mobile robotics. A line scanner is a laser range finder, which is swept through an arc by a spinning mirror. The sensor detects the return from the laser and calculates distance from time of flight at discrete angular increments around the disk so described. A line scanner oriented so the plane of detected points is horizontal (e.g., the axis about which the mirror spins is vertical and the angle between the mirror and its axis is 45°) is useful in real-world terrain where objects of interest (walls, etc.) are also vertical, such as an indoor environment. This sort of sensor is insufficient for general obstacle detection and terrain mapping but can be used to generate a useful first approximation, as vertical terrain features tend to be the most salient.

A line-scanning unit built by the German company SICK AG was used on many of the mobile robots competing in the recent Defense Advanced Research Projects Agency Grand Challenges for autonomous unmanned ground vehicles. The SICK unit, however, is too large and heavy for this application. Instead, a device similar to the principal geometry sensor was used.

A smaller line-scanning device, the URG-04LX [5], is available from Hokuyo. This sensor works very much like the SICK scanner but is small (2 x 2 x 3 in) and lightweight (165 g). It sweeps an arc of 240° at a rate of 10 Hz, returning range measurements at intervals of 0.36°. This corresponds to approximately one data point per inch at the maximum range of 4 m and 683 data points every 100 ms to process to maintain real time. The URG sensor is mounted to the frame of the robot so that the plane of the measurements is horizontal and at the height of the robot. This is consistent with using the sensor to avoid right prismatic (cuboid) obstacles, and it also enables the mapping of indoor terrain, which is predominantly bounded by vertical planes.

A second line scanner is mounted at the front of the robot, directed at the ground ~1 m ahead of the robot. This scanner senses the terrain the robot is just about to drive onto. The horizontal line scanner receives no sensed data from the ground, so the second scanner is depended upon to assure that there is indeed ground to drive upon and that the terrain is smooth enough for the robot to traverse. Ideally, this sensor would look out 3 m ahead so there would be time to stop if, for example, the sensor detected the top step of a flight of stairs. The look-ahead distance may be changed as researchers gain experience with the system.

3.1.2.2.2. Imaging Ladar

More detail concerning the geometry of the environs is available from an imaging ladar sensor. Such a sensor acquires range data as a set of range vectors centered at a focal point and organized as an image, e.g., rows and columns of data points. Surveying ladars, such as those available from Riegl USA, Inc., provide high-resolution three-dimensional data at ranges over 100 m, but the range measurements are sequential and too slow for mobile applications. Ladars built specifically for mobility applications, such as the product built by General Dynamics Robotic Systems for the Army's Autonomous Navigation System, are substantially faster, but today's technology is too large and heavy for use on this small robot.

A recent technology known as "flash ladar" is based on camera technology, enabling fast data acquisition as well as light weight and compact size. Such a device illuminates the environment with light modulated at a known frequency and determines time of flight from the phase shift of the reflected energy incident on each pixel imager. Devices using this technology are available from PMD Technologies GmbH, Mesa Imaging AG, and possibly others.

The device selected for the ATRV sensor upgrade is the Mesa Imaging SwissRanger SR-3 000 [*6*]. This sensor collects frames of range data 176 x 144 pixels at a rate of 30 Hz over a field of view of 47.5° x 3 9.6° (0.27° per pixel). Maximum range is advertised as 7.5 m, limited by the nonambiguity constraints of the measurement technique, but several papers in the literature indicate a shorter useful range. Range resolution is specified by the data sheet as 1% of range. A cursory evaluation of the sensor revealed a sensitivity to bright lights, resulting in washed-out regions of the image, which must be further investigated.

The SwissRanger is mounted on an existing pan-tilt unit on the deck of the ATRV. The pan will be used to compensate for the narrow fields of view of the various sensors mounted on the unit. The tilt axis will likely be set at a fixed look-down angle, which provides a "good" amount of information about the ground immediately ahead of the robot while not sacrificing too much information about overhanging objects.

3.1.2.2.3. Stereo Vision

There will be times when it is necessary to sense geometry at ranges greater than that provided by the active sensors. Computer-based stereo vision can provide range images at distances of tens of meters, but it has been seldom utilized outside the laboratory (and in planetary exploration). In part, this is because the sensors (conventional cameras) are inexpensive, but the computing to process the camera images into a range image was "do it yourself"—the phenomenon was well understood and algorithms were widely available, but there was no integrated stereo "system" delivering range images.

The recent availability of "Stereo on a Chip," from Videre Design LLC, has changed the maximum range available from an active sensor [7]. A field programmable gated array packaged with the complementary metal oxide semiconductor video imagers computes disparity (a function of range) at each pixel of the VGA-resolution image at 30 Hz, reducing the workload of the host processor substantially. The 4.5-mm lens images a field of view of 59° horizontal x 41° vertical. As configured for UVTD's application, the range resolution at 8 m is computed by Videre Design's online calculator to be roughly 2.5 in; ranges closer than 0.5 m are unavailable. The stereo sensor is expected to deliver a dense sampling of a (possibly imprecise) range function, allowing ranges to be estimated beyond the ability of the ladar sensors. Reflectance values from the stereo system are also available on a frame-by-frame basis, enabling data integration and/or fusion.

The stereo sensor will be mounted on the pan-tilt unit near the SwissRanger so the region sampled by both sensors overlaps. The overlap among the fields of view of the various sensors, shown in figures 3 and 4, will be exploited in any way possible.

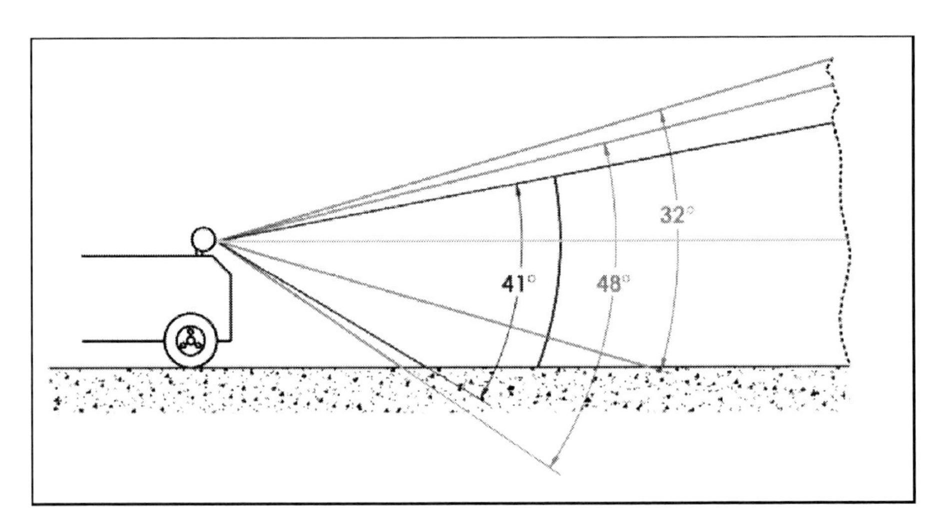

Figure 3. Sensor fields of view are shown here in side view, according to the following key: stereo in blue, color camera in green, flash ladar in red, horizontal line scanner in magenta.

3.1.2.3. Inertial Reference

All the sensors described report their sensed data in the coordinate frame of the sensor itself. In order to integrate the information over time, it is necessary to transform the data to a common coordinate system, preferably a world-fixed coordinate system. The conventional way to do this is to sense the robot location from a sensor such as GPS and the robot orientation from a compass, and augment these sensors with time derivatives of each from an inertial reference sensor suite and odometry. In the case of this ATRV, the ideal operational area is where GPS is unavailable and where compass readings may be compromised (and possibly in unmapped regions). In this case, high-quality time derivatives of position and orientation are wanted because of the integrations required.

A 3DM-GX1 inertial reference sensor (IRS) from MicroStrain, Inc., [*8*] provides orientation and acceleration for the upgraded robot. Raw data from embedded accelerometer, gyros, and magnetometers are fused by the sensor itself. Alternative IRSs are available but were not considered since sensors from the MicroStrain product line are used on other division assets, and performance was deemed acceptable.

3.1.3. Power

Power for the robot as a whole was left unchanged from the original ATRV, that is, dual 12-V deep-draw batteries with off-board recharging. It remains to be seen whether the duration available will be sufficient for research missions. Power for peripherals was shifted from the computer power supply of the original ATRV power architecture to a custom power distribution board supplying regulated 5 and 12 V through bussed terminal strips.

The CPU of the original ATRV computer was rated at ~30 W, while the CPU selected for the upgrade was rated at 125 W, so the computer power supply was upgraded as well. The onboard power supply, the M2-ATX 160W [*9*], was selected based on its tolerance of a wide range of input voltages (6–24 V) and its form factor, which corresponds to the dimensions of the case selected. The maximum input operating voltage, outlined in the specifications, is 24 V, which is marginal for a battery system consisting of two 12-V batteries. A more recent release, the M2- ATX-HV, claims an even higher maximum input voltage and would be a more conservative selection.

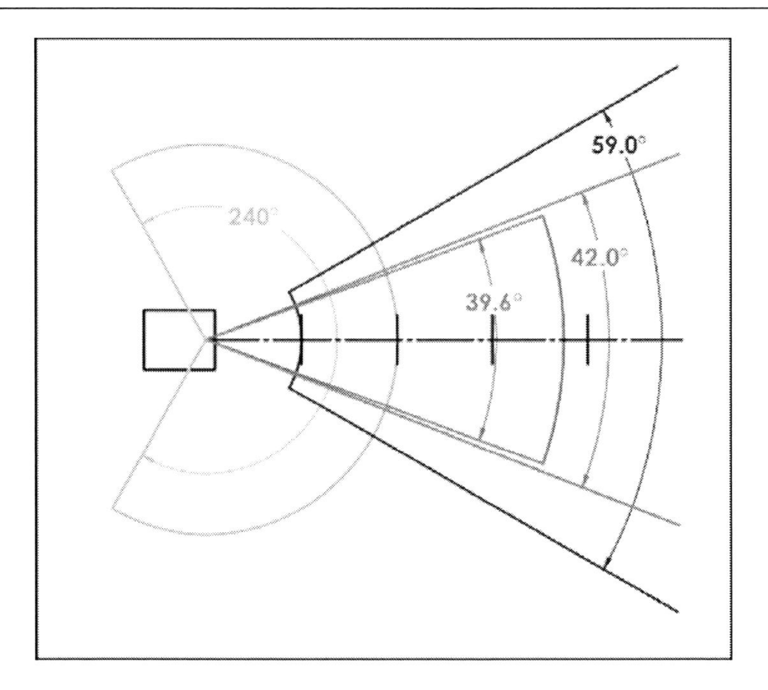

Figure 4. Sensor fields of view are shown here in top view, according to the following key: stereo in blue, color camera in green, flash ladar in red, horizontal line scanner in magenta. Black centerline shows crosshatches at 2-m intervals out to 10 m.

3.1.4. Motor Control Board

The motor control board was replaced as well, though the two wheelchair motors and gearboxes driving the ATRV's four wheels were retained. The AX3 500 product from Roboteq [*10*] was selected for the following reasons:

1. It is designed to run with 24-V rechargeable batteries.
2. It supplies maximum motor current of 40 A for each of two brush-type electric motors, well beyond the 11-A rating of the wheelchair motors.
3. Input to the motor control board through RS-232 is available. A translational velocity /rotational velocity command is native.
4. The board supports optical encoders, also part of the original ATRV motor suite. The encoders enable closed-loop velocity control, an essential element for control of a skid- steer vehicle where soil resistance is uncertain. In addition, the encoder counts can be monitored through the serial link, providing an odometry function.

5. It has two distinct safety shutoff modes, including one which can be easily asserted from a remote radio control (RC) controller.

While each of the elements listed are essential to the application, a number of other features bring added value to the control board, notably the ability to control the robot from an RC remote control. This capability is very useful in logistic operations, such as maneuvering the robot into its parking place at the end of the work day.

3.1.5. Safety Circuitry

The original ATRV was equipped with four e-stop mushroom buttons on the robot. To stop the robot in case of a software failure, it was necessary to approach the robot and push one of the buttons. One of the goals of the upgrade was to increase the safety of the robot by enabling a software-independent kill capability from a distance, so the robot can be brought to a halt without jeopardizing the operator.

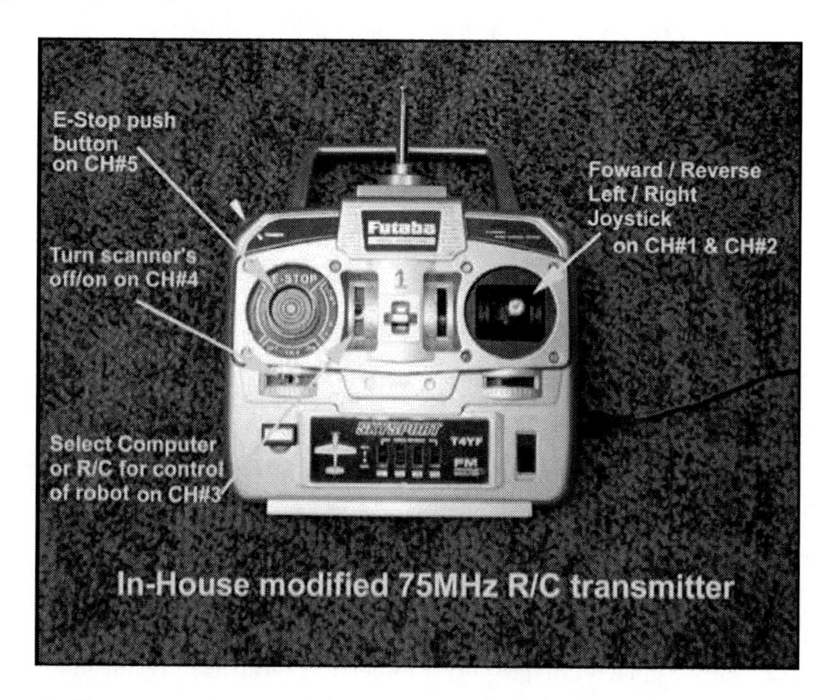

Figure 5. The remote control provides the operator the ability to stop the robot from a safe distance in case of emergency and operate it manually with the joystick.

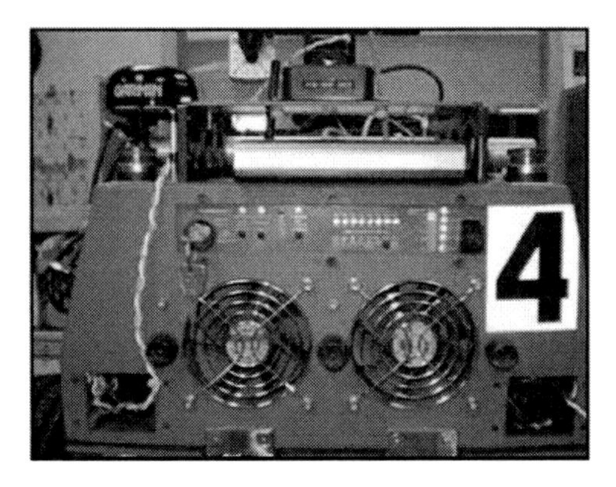

Figure 6. Display panel at ATRV rear.

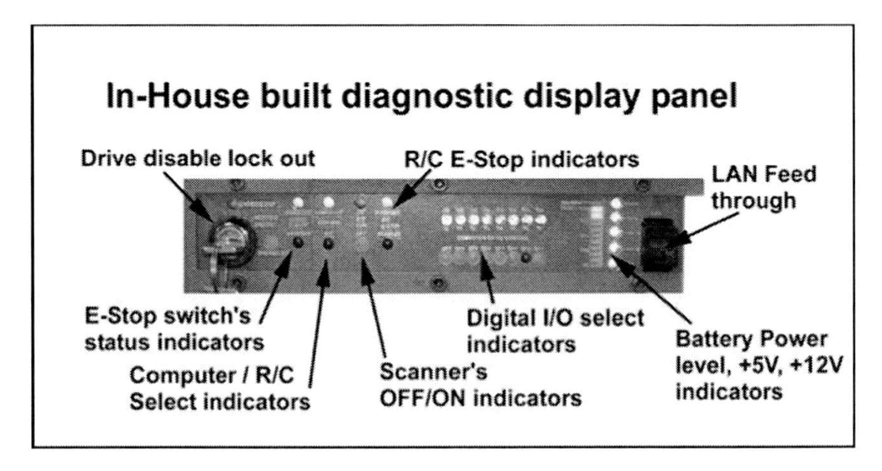

Figure 7. The display panel shows the status of a number of internal states.

Using a safety mode suggested by the manufacturer of the motor control board, there are now three means of stopping a runaway robot. The first is by means of the e-stop buttons on the robot. These cause an e-stop input on the motor control board to be activated. The second is a red e-stop button on the remote control. Pressing this button actuates the same input through one channel of the RC radio. Both require a reset from a key-switch on the robot body before motor control is restored.

The third mode stops the robot by a new mechanism. The motor controller can accept commands from either the serial link connected to the onboard computer or from the RC receiver linked by a dedicated radio channel to the remote control. The remote control determines which signals reach the control board input by means of an electromechanical relay.

The default control is from the remote control, and it must be ceded to the computer by a manual switch on the remote control. If the program running on the computer is judged by the operator to be in dangerous error, the operator can throw the switch on the remote control, which seizes control from the computer and returns it to the remote control in the hands of the operator. This safety paradigm is similar to that used on UVTD unmanned air assets. Figure 5 depicts the operator's remote operation and safety control. Figures 6 and 7 illustrate status monitors, allowing the operator to confirm elements of the robot control state.

3.2. Robot Software

Since the original software provided by RWII for the ATRV Jr. was proprietary, replacing the computers meant replacing the operating system and supporting software (including the device drivers). Thus the decision was made to utilize the Open Source robotics package Player (from the Player/Stage project [*11*]), which provides a convenient hardware abstraction layer and a multitude of popular robotics device drivers. Player is also used by numerous academic institutions with robotics programs including the University of Pennsylvania, Georgia Institute of Technology, and the University of Southern California [*12*].

3.2.1. Organization

The software on the robot computers is divided into three layers: base, core, and brain (see figure 8).

The base layer contains relatively static library code and the core operating system facilities that are common across all machines, including a properly configured kernel for the CPU architecture. Debian [*13*] GNU/Linux is the operating system of choice, primarily to gain the benefit of its package management system Advanced Packaging Tool (APT) as well as the ease of configuring a custom system using the very useful debootstrap tool.

The core layer is an abstraction layer providing higher-level functionality to the layer above it. It contains the Player system and any shared libraries required for robotic development, including the agent architecture under development (see section 3.4). The ATRV Jr. currently under development has a custom driver plug-in for Player that was developed for the Roboteq controller, as well as the libraries required for the SR3 000 Flash Ladar. In general, more volatile libraries (rapidly changing open source and internal packages) will reside in this layer.

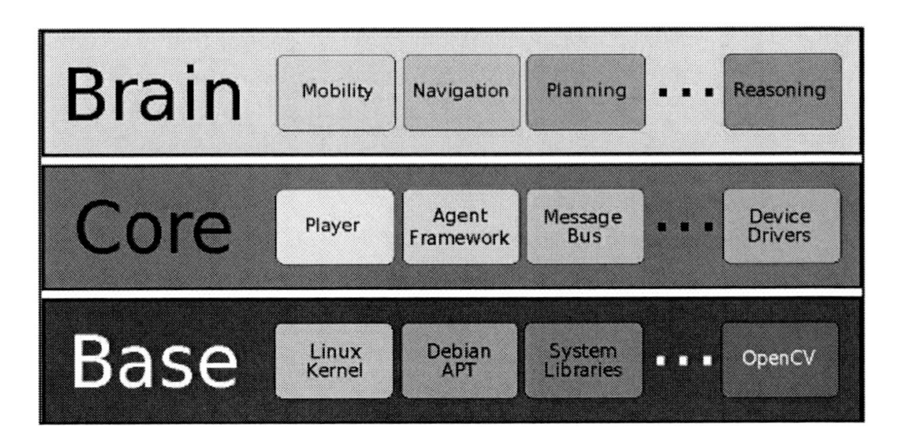

Figure 8. High-level software organization. The smaller items in each level indicate the types of components in each layer.

The brain layer is the effective application layer where agents and high-level behaviors can be implemented. Thus it will contain the custom scripts, executables, and data that compose the actual behavior of the robot. Currently, the brain layer is not implemented in the upgrade; however, work is underway to address that problem (section 3.4).

3.2.2. Operating System

The base operating system was constructed to be relatively small and boot fast. Debian GNU/Linux, however, provides an installation script that downloads a minimal Debian operating system without the Linux kernel. The minimal system includes only a small subset of a typical GNU/Linux distribution, so scripts were created to add in additional software readily available from the Debian software repository.

The omission of the Linux kernel from the minimal system is deliberate; more than likely, someone building a custom distribution will want to custom configure a kernel for specific hardware, as is the case for this project. Thus the scripts choose among several custom kernel configurations based on a given keyword, build the kernel, add it to the system base directory, and build a disk image that can be copied directly to the robot's hard disk.

The current custom build of Debian is around 300 MB, which includes all the required kernel modules, base libraries, and extra libraries needed to comfortably support the Player system and most anything else needed (this includes the Debian-provided version of the OpenCV computer vision library). That is considerably larger than the initial goal and is mostly the result of some extraneous dependencies on GTK+ libraries within Player, which can be removed when time permits. However, the system does boot in just under 10 s, which is quite good. Boot time can be improved by optimizing the operating system.

3.2.3. Device Abstraction

As mentioned previously, the Player system is a "robot device interface and server" and acts as a device abstraction layer to elements of the robotic architecture residing in the core and brain layers. Since Player is now one of the most popular open source robotics libraries, it already has support for many of the devices used, including IEEE 1394 cameras, USB cameras, the SR3000 flash ladar, and the Microstrain inertial measurement unit (IMU). Adding a device is straightforward—pick or develop a Player interface that defines an abstract representation of a device (e.g., the laser interface defines how to talk to a laser-ranging sensor without worrying about the particulars of device initialization, configuration, or communications protocol). Most drivers then provide a specific implementation of an existing interface and a configuration file format for specifying hardware-specific parameters in a runtime-configurable format.

The ATRV Jr. currently has one custom device driver implementation for the Roboteq motor controller, described in the next section.

3.2.3.1. Roboteq Device Driver

The Roboteq device driver written by UVTD implements the Player position2d interface, which can be used to control planar mobile robots. The interface provides the facility to issue velocity commands (x_dot, y_dot, theta_dot), position commands (x,y,theta), speed/heading commands (v,theta), and car commands (v,theta), which the driver may ignore or implement according to the platform configuration. The current version of the Roboteq driver only understands the velocity commands and assumes the presence of encoders and the use of mixed mode, closed-loop serial operation to the actual motor controller device.

3.2.3.2. Player Configuration

Player is very flexible and does not assume or impose much on a system design. The core of Player is the device abstraction, but Player also provides a Transmission Control Protocol (TCP)-based server that allows multiple remote connections and controls for each device configured for that server. In most cases, there is one Player server per robot providing a connection to all the devices configured for that robot. However, the server is not implemented in a concurrent manner; therefore, the configuration in use on the ATRV Jr. takes advantage of multiple CPUs by providing one Player "server" per robotic device, e.g., a stereo vision server, an IMU server, etc. This provides the same abstraction as one server for all but allows the servers to run concurrently (and therefore block concurrently, if need be).

3.3. Development Environment

Building an essentially new robotic system from the ground up[8] requires that configuration management (CM) and software engineering issues be addressed. Section 3.3.1 highlights the motivation and derived requirements that guide the design of the environment described in section 3.3.2.

3.3.1. Motivation and Requirements

Creating a formal CM environment is motivated by a desire to do the following:

- Keep the robot systems clean and free from version incompatibilities.
- Allow new engineers/developers to become productive with the available tools.
- Facilitate access to the code.
- Share the maintenance and development of the system across the set of contributors.
- Support more than one robot system.
- Encourage and/or enforce compliance with software engineering practices in order to improve code quality.

[8] At least from the software point of view.

These goals were used to define the following high-level requirements:

1. The robotic system software must be versioned from a central server.
2. The development model will be a host-target configuration.
3. The development systems must have a common set of libraries and tools, and therefore be imaged from a central server.
4. The development systems must have access to the robot system.
5. The central server and development systems must reside on the same network.
6. The central server must provide reliable data storage.
7. At least two laptops must be available for operating, debugging, and logging data from the robots.
8. It must be easy to update the robots with newly developed software.

The system design that implements these requirements is detailed in section 3.3.2. However, two particularly important ramifications of this CM warrant further description in the following sections.

3.3.1.1. Providing a Clean Slate

Since the ATRV Jr. is a shared resource that will be utilized by multiple researchers often investigating somewhat orthogonal topics, providing a clean operating environment is essential. Extraneous software should be kept to a minimum, with an eye toward the essentials that make the robot run. Not only does this leave more storage capacity, but it speeds up the system and reduces the chance that software might conflict with mission-critical functions. While installing a programmer's text editor and all the development libraries seems harmless, it also seems completely extraneous for a robot running a real mission[9]. In addition, the host-target model helps to ensure that the robot does not get out of sync with the repository. Facilities will be provided to build the relevant portions of the system (i.e., base, core, brain) and then transfer those portions to the robot (which enables a "set it and forget it" behavior).

[9] Recently, there has been some discussion on relaxing these restrictions through careful build-time parameters; e.g., a build switch could indicate whether the robot is being used for development or "production" use.

3.3.1.2. Enabling Software Reuse

The ATRV Jr. upgrade is seen as an opportunity to begin the construction of a development environment and software platform conducive to creating state-of- the-art robotic vehicles based on the x86 architecture. Part of accomplishing this goal is providing for effective software reuse. Thus this approach relies on system connectivity, redundant data storage and automated backups, capable version control and a defined usage policy, and modular software design. Systems need to be connected in order to facilitate source code-level sharing and allow the systems to enforce a check-in policy. Reliable data storage is a requirement to prevent loss of valuable work and knowledge. Version control is a must in order to provide a well-known repository of code and a means to keep it organized and safely shareable. Finally, modular software design (see section 3.2.1) is the real key to reuse; while it is clear that a given robot will have some customized pieces of software that are likely unusable in a different robot, it is equally clear that many algorithms, frameworks, and sensor device drivers can be used across many robots. This is ultimately the purpose in open source software packages like the Player/Stage project and Yet Another Robot Platform (YARP) [*14*]. A very conscious decision was made in this project to consider the packages available and reuse others' hard work as much as possible (i.e., the use of Player/Stage and the Unified System for Automation and Robot Simulation [USARSim] [*15*] as well as the ideas, algorithms, and/or code from packages like YARP and the Mobility Open Architecture Simulation and Tools [MOAST] [*16*] framework). This has, without a doubt, accelerated the ATRV Jr.'s software development significantly.

3.3.2. Environment Design

Based on the requirements stated previously, the environment consists of the following elements: an online Ubuntu GNU/Linux workstation, an offline network, a development server, multiple development laptops, robots, a version control system (VCS), a remote synchronization server, and a host of scripts implementing build and update functionality. A high-level diagram of the topology is shown in figure 9.

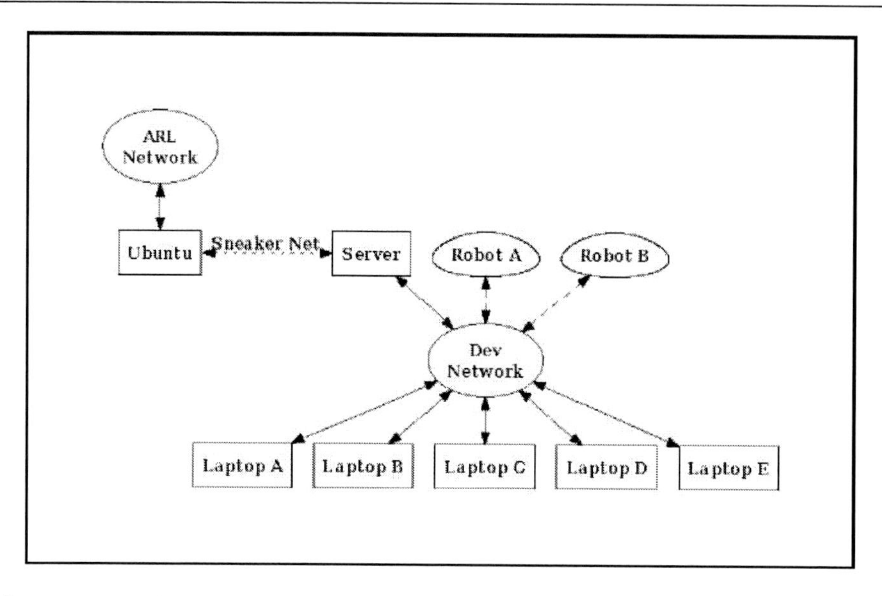

Figure 9. An Ubuntu workstation is online and provides updates to the development server through a manual process, which, in turn, provides automatic updates for the offline network.

3.3.2.1. Online Ubuntu Workstation

Since Ubuntu [*17*] GNU/Linux is based on Debian GNU/Linux, it inherits the excellent APT and an extensive repository of software[10]. The online Ubuntu workstation provides access to this resource, and custom scripts generate local repository mirrors that can be transferred to the offline development server.

3.3.2.2. Offline Network

An offline network is necessary since the custom robot systems (based on Debian GNU/Linux) and development systems (based on Ubuntu GNU/Linux) cannot currently gain access to the Internet[11]. This network connects the development server and workstations together to provide version control and system software updates, and also allows the workstations to connect to the robots.

---

[10] Often, software is not available anywhere else without a manual installation procedure: find dependencies, download, compile, and install everything in the proper order (a time-consuming process).

[11] For reasons beyond the scope of this report.

3.3.2.3. Development Server

The development server is an Ubuntu GNU/Linux-based system that provides the master VCS repository (on a RAID-1[12] setup providing reliable data storage) as well as all the required libraries and tools for software development. The development server also provides a file synchronization utility (rdist[13]) service for the development laptops and considerable computational power for robot simulations.

3.3.2.4. Development Laptops

The development laptops are intended to be shared resources that serve as both workstations for software development and control stations for system testing. They are configured to enforce a VCS check-in schedule in order to ensure little or no data loss should some hardware fail. The development server automatically provides remote software updates to each laptop; thus all laptops have synchronized file systems. At least two of these laptops can be used specifically as control stations during testing or demonstration.

3.3.2.5. Version Control System

An important part of a basic software development environment is always the VCS. It stores the history of the project and all development artifacts (i.e., it doesn't need to hold only source code). To offer the greatest flexibility, a distributed VCS[14] is utilized. This allows easier branching and merging of individual lines of development (i.e., facilitates experimentation as well as bug fixes for stable releases) and makes it possible to develop outside the network. Since any "clone" of the repository is actually a full repository in itself[15], one can transfer the repository to a completely different network or simply take a repository offline for development away from the server. When this "disconnected" repository needs to synchronize with the "connected" repositories, the differences are automatically merged (whether through a network connection or the exchange of patch sets through some medium like e-mail). This configuration

---

[12] Raid = redundant array of independent disks.

[13] Remote file distribution: a method of distributing software updates from a central location to multiple remote sites. The behavior is implemented with a client (rdist) and a server (rdistd) (*18*).

[14] Currently, Mercurial is the distributed VCS of choice. However, Git is also being investigated for its power and flexibility.

[15] Contrast this with centralized, nondistributed VCSs like CVS and Subversion (one must have access to the server in order to check in code).

allows for the most flexibility, especially for contributors that may not have access to the network.

3.3.2.6. Remote Synchronization System

As mentioned in section 3.3.2.3, the development server hosts a rdist daemon that ensures the file systems across the server and laptops are identical. Note that home directories are not synchronized, while all the important development tools and libraries are kept up to date. This does require some manual updates to be made to the development server itself (see section 3.3.2.2) but only when changes are required (security update or new version of a library, etc.).

3.3.2.7. Build Scripts

A library of build scripts is distributed as part of the robot software distribution's development repository. These build scripts automate and therefore simplify a large portion of the effort needed to construct aspects of the robot software package. For example, there are scripts to download and build the base component as well as scripts to automatically configure the Player library for the ATRV Jr.'s specific device configuration. These scripts are expected to be developed over time to include the most configurable aspects of the system in such a way as to greatly facilitate the construction of a software package for an entirely new piece of robot hardware.

3.3.2.8. On the Host-Target Development Model

The development environment for the new ATRV Jr. is based on a host-target environment, meaning a large portion of development happens on a developer's workstation or a stand-in platform, and the resulting product is transferred to the robot for testing. While this is a departure from the previous ATRV configuration that allowed self-hosted development, it is not as bad as it seems. By using the Player framework for hardware device abstraction, researchers get to make use of supported simulation environments, including (but not necessarily limited to) Stage, Gazebo, and USARSim. A debugger is present in the default base distribution and the host-target separation can be relaxed for situations where it makes more sense to develop directly on the robot (however, review section 3.3.1.1 for reasons to avoid that scenario).

3.4. Toward a Robotic Agent Architecture

While the software support on the ATRV Jr. can, at present, be used for application development and provides a high-level division between modules responsible for different tasks, it does not address how individual components work together to achieve the total behavior for the system. The previous ATRV Jr. software architecture is similar to the current one in that it is based on a collection of services using the Common Object Request Broker Architecture. While services can go a long way in providing reusable device abstractions and behaviors, the architecture concepts discussed in the next sections extend that paradigm to a higher level—that of cooperating agents.

3.4.1. Issues

In the context of developing a framework to control the ATRV Jr. platform, the term architecture refers to how the overall behavior, intelligence, and control of the ATRV will be arranged in software. The term "agent" refers to a self-contained entity that implements a perceive -> think -> act loop and can communicate with other agents. The idea of the agent architecture is to implement the "brain" as a collection of loosely coupled agents acting individually to perform specific tasks and concurrently acting together to enable the emergent behavior, intelligence, and control of the robot. Since an agent can be thought of as a superset of a service, an agent-based architecture can parallelize and distribute across multiple CPUs or systems as well as service architectures can. An agent-based architecture also distributes better than a monolithic system (which cannot run across multiple systems at all).

The concept of agents is not particularly new or unique. Many robot software implementations provide the capability to construct stand-alone components that can communicate with other components, sometimes within a prescribed methodology, sometimes without any guidelines for application structure at all, and almost always tied to a specific implementation language. The following questions are addressed with a new architecture design based on agents:

- What infrastructure and performance capabilities are required to effectively support advanced perception research?
- What infrastructure is required to support autonomous behavior research?

- Is there a way to integrate concepts from the cognitive artificial intelligence perspective into a cohesive software framework that can also include more primitive capabilities?
- Can the architecture be efficient enough to run aboard relatively small robots but flexible enough to support more advanced software designs (i.e., distribution, concurrency, clustering)?
- Is there a way to integrate existing software as a functioning component within the architecture?

3.4.2. Initial Design Topics

The architecture is broken down at the highest level into two components: the infrastructure and the interfaces. There really is a third component, the actual implementation of the agents, but it is orthogonal to the architecture design. Note that hardware abstraction is handled by the Player software and is therefore omitted in the following discussion.

3.4.2.1. Infrastructure

The infrastructure component includes everything that enables the agents to cooperate and perform their functions but includes and prescribes nothing related to actual robot behavior. In other words, it acts as a substrate that supports the agents both during development and runtime.

Initial requirements for the infrastructure include:

- be as lightweight as possible
- enable multilanguage agent implementations
- support appropriate peer-to-peer and service-oriented constructs
    - discovery and lookup
    - group communication/multicast
- utilize message-passing abstractions for all inter-agent communication
- allow prioritized messages to disambiguate conflicting requests
- provide efficient implementation constructs where possible
- provide a simple method for launching agents

3.4.2.2. Interfaces

The interfaces describe the kinds of agents or services one uses to build a robotic intelligence system (but not necessarily how to build the agents). By specifying interfaces, one can effectively describe the requirements of an agent without specifying implementation details and decoupling agents from other agent implementations, providing for a more fluid system design that allows for parallelism, distribution, multilanguage support, and pluggable algorithms. For example, table 1 lists some agents and services researchers would like to implement on the ATRV Jr.

Table 1. A representative example of agents and services for the ATRV Jr., roughly organized by complexity.

Low	Middle	High
Safety	Identification	Planning
Mobility	Tracking	Task
Mapping	Telemetry/reporting	Health
Navigation	Manipulation	Cooperation
Geometry	Memory	Metareasoning
Obstacle	Context	Human interaction

3.4.3. Existing Work

The following sections provide brief surveys of a number of existing robot software systems that may serve as foundations or guidelines for future work.

3.4.3.1. Player

Player provides a multiclient/server paradigm over TCP for interacting with robot hardware and does not impose or suggest any other structure or organization. The common use case involves a single server representing the devices on a robot and one or more client programs implementing behavior. Player is written in C/C++ and provides C, C++, and Python client interfaces directly, while others have provided a host of interfaces for other languages (e.g., Java, Octave, Matlab). Refer to section 3.2 for information on the way this project is already leveraging the capabilities of this software package.

3.4.3.2. The Mobility Open Architecture Simulation and Tools (MOAST)

MOAST framework is based on the four-dimensional real-time control system architecture developed at the National Institute of Standards and Technology [*19*]. The framework divides functionality into vertical hierarchies called echelons (e.g., primitive echelon, autonomous mobility echelon), where each echelon is further divided into functional components (e.g., sensor processing, world modeling). The Neutral Messaging Language is used for platform-independent communication between modules (which can be distributed across different systems). Like most of the systems described here, MOAST is open source.

MOAST is one of the most promising candidates for integration into this project if further investigation indicates it directly supports or allows development of the needed infrastructure and interfaces outlined in section 3.4.2.

3.4.3.3. Coupled-Layer Architecture for Robotic Autonomy (CLARAty)

The CLARAty project is described as a reusable robotic software framework. While not truly open source, this project has many algorithm implementations that may be useful as standalone elements in a new system. Integration may be difficult, however, since while the design is very modular, it does not seem to support concurrency or distribution explicitly.

3.4.3.4. Other Packages

While investigating previous work, the authors have come across a large number of software packages aimed at developing mobile robots, including but not limited to YARP, RobotCub, Saphira, OROCOS, MARIE, FlowDesigner, and RobotFlow. Descriptions of these packages are outside the scope of this report; a future report describing the proposed agent architectures will provide more in-depth discussion of the state of the art.

4. Conclusion

The ATRV, even with its upgrades, is not by any means a military robot. It is a platform that provides infrastructure (mobility, power, communication) supporting the research elements of the program, i.e., perception processing and autonomous behaviors. As these behaviors take shape and mature, they will be transitioned to robots designed for the field, along with sensors, platforms, and computing elements coming from other efforts.

The upgraded ATRV has computing power and communications at the state-of-today's practice, a suite of sensors with a variety of technologies and complementary strengths, an architecture built on the foundations of the best robotics research institutions in academe, and safety substantially improved over the baseline. Lessons have been learned in the upgrade, which strengthened the skills and understanding of the researchers involved. It is to be expected that the process, as well as the product, of the upgrade effort will enhance the research efforts of ARL's Unmanned Systems Division for years to come.

Appendix. Sensor Selection for a Fast Indoor Robot

This appendix describes mobility sensor requirements for a robot capable of speed comparable to that of a human in similar circumstances. The assumed domain is indoors or mild outdoor terrain. The implication of indoor terrain is that there is little room for evasive action (a hallway might end in stairs that span the entire width of the hallway), so the most critical response to a hazard is to stop. Sensors must be able to detect an incipient hazard at a range sufficient to allow room for a complete stop.

The speed at which a human moves indoors can be parameterized as follows. Human walking speed is roughly 2 m/s, while human running speed can be roughly bounded by the rate of a world-class sprinter, say 10 m/s. Indoors, it seems unlikely that a human will move at a sprinter's pace, so assume a maximum speed of 5 m/s.

The deceleration of the robot to a complete stop is governed by brakes (for a vehicle-like robot) and coefficient of friction. For design purposes, the coefficient of friction is assumed to be 0.5, limiting deceleration to roughly 5 m/s/s.

The distance required for the robot to stop can now be calculated. From 5 m/s, with braking acceleration limited by the coefficient of friction, it will take the robot 1 s to decelerate to a standstill, during which time it will cover a 2.5-m distance. The implication for the sensor suite is that the robot must be able to reliably detect obstacles at a distance of at least 2.5 m. A safety margin for reaction/response time should also be included. A human can react in 0.1 s; if a robot reacts in the same time (not necessarily a conservative assumption), an additional 0.5 m must be added to the sensor detection range for a total of 3 m.

In general, sensors have an easier time detecting a positive (above the local surface) than a negative (below the local surface, e.g., a hole or depression) obstacle. However, both kinds of obstacles must be detected.

References

[1] VIA Technologies, Inc., Mini-ITX announcement. http://via.com.tw/en/initiatives/spearhead /mini-itx/ (accessed 1 September 2008).

[2] Alfa Network, AWAP608 product page. http://dplanet.biz/alfa.com/product_info.php ?cPath=1 59_33_55&products_id=1 56 (accessed 3 September 2008).

[3] Unibrain, Fire-i digital camera product page. http://www.unibrain.com/Products/VisionImg /Fire_i_DC.htm (accessed 27 August 2008).

[4] Maxbotix, LV-MaxSonar-WR1 Beam Plots. http://www.maxbotix.com/Performance_Data .html (accessed 27 August 2008).

[5] Acroname Robotics, URG-04LX product page. http://www.acroname.com/robotics/parts /R283-HOKUYO-LASER1 .pdf (accessed 27 August 2008).

[6] Mesa Imaging, SR3000 product page. http://www.mesa-imaging.ch/pdf/SR3000_Flyer _Sept07.pdf (accessed 27 August 2008).

[7] Videre Design, STOC product page. http://www.videredesign.com/vision/stoc.htm (accessed 27 August 2008).

[8] MicroStrain, Inc. 3DM-GX1 product page. http://www.microstrain.com/3dm-gx1.aspx (accessed 27 August 2008).

[9] Mp3Car.com, M2-ATX product page. http://store.mp3car.com/M2_ATX_160W_Intelligent _DC_DC_PSU_p/pwr-0 15 .htm (accessed 27 August 2008).

[10] Roboteq, AX3500 product page. http://www.roboteq.com/ax3500-folder.html (accessed 27 August 2008).

[11] Gerkey, B.; Vaughan, R. T.; Howard, A. The Player/Stage Project: Tools for Multi-Robot and Distributed Sensor Systems. *Proceedings of the 11th International Conference on Advanced Robotics*, Coimbra, Portugal, June 2003; pp 3 17–323.

[12] Player Wikipedia, Player Users Wiki page. http://playerstage.sourceforge.net/wiki /PlayerUsers (accessed 1 September 2008).

[13] Debian Home Page. http://www.debian.org/ (accessed 3 September 2008).

[14] YARP Home Page. http://eris.liralab.it/yarpdoc/index.html (accessed 1 September 2008).

[15] SourceForge.net, USARSim description. http://sourceforge.net/projects/usarsim (accessed 1 September 2008).
[16] SourceForge.net, MOAST description. http://sourceforge.net/projects/moast/ (accessed 1 September 2008).
[17] Ubuntu Home Page. http://www.ubuntu.com/ (accessed 3 September 2008).
[18] MagniComp, Rdist Home Page. http://www.magnicomp.com/rdist/ (accessed September 2008).
[19] Scrapper, C.; Balakirsky, S.; Messina, E. MOAST and USARSim: A Combined Framework for the Development and Testing of Autonomous Systems. *Proceedings of the SPIE Defense and Security Symposium*, Orlando, FL, April 2006.

In: New Robotics Research
Editors: E.D. Wagner et al, pp. 63-76
ISBN: 978-1-60741-093-5

Chapter 4

POLYMER MATERIALS FOR GROUND MOBILE MILLIMETER-SCALE ROBOTICS*

Ryan Rudy, Ronald G. Polcawich and Jeff Pulskamp

Abstract

This project is closely tied with the ongoing work of visiting Professor Kenn Oldham and the U.S. Army Research Laboratory's (ARL) joint effort on creating highly flexible, large payload capacity joints for a ground mobile millimeter-scale robot. The fabrication process to add parylene coatings to the piezo-microelectromechanical systems (piezoMEMS) actuator process has been characterized using test structures. Scanning electron and optical microscopy of the joint assemblies; analysis of the coating technology for trench fill; process robustness to exposure to solvents and photolithographic processing; and adhesion of parylene to both platinum and lead zirconate titanate (PZT) thin films have been completed on two separate fabrication sequences. Parylene coatings have been successfully applied to both platinum and PZT thin films and the challenges associated with parylene survival with multiple fabrication process steps have been evaluated. Future work will include full release of test structures on the existing wafers in fabrication as well as implementation of process improvements into a fully functional piezoMEMS plus parylene actuator joint.

* Excerpted from Army Research Laboratory Report, ARL-TR-4659, Updated December 2008

1. Introduction/Background

There has recently been a push within the U.S. Army Research Laboratory (ARL) to create ground mobile millimeter-scale robotics for surveillance purposes. Current challenges are faced in fabrication due to the relatively large amount of weight the base structure must support. This weight is due to power systems, control systems, and communications systems as well as any other additional components, as seen in the conceptual design in figure 1 [1].

In order to address this issue, high aspect ratio leg links will be fabricated to increase the weight bearing capabilities. Previous work has been done in designing and processing such devices with a protective parylene skeleton as well as flexible parylene joints. Parylene was chosen because it can provide uniform coatings within deep narrow spaces, allowing for the coating of large aspect ratio trenches for the leg links. Parylene also provides the flexibility needed to create leg links with a large range of motion [2].

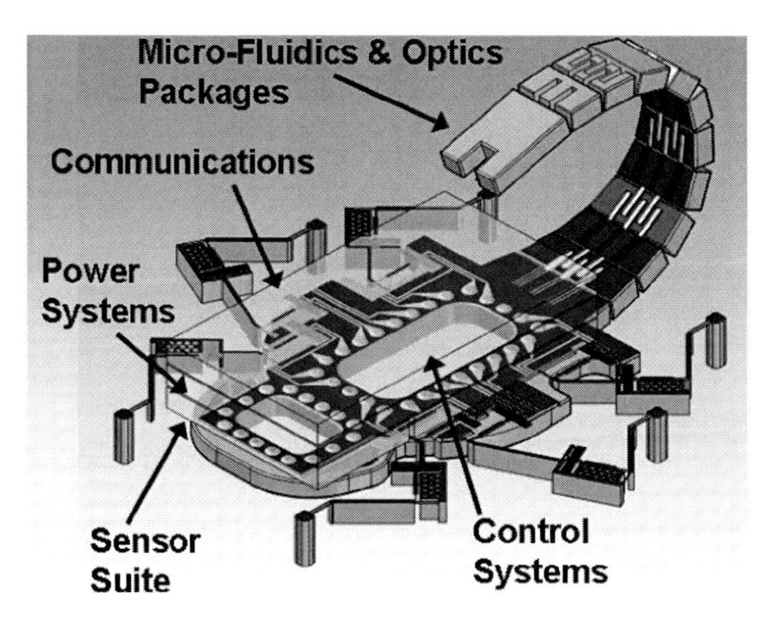

Figure 1. The conceptual design of the ground mobile millimeter-scale robot highlighting various components and systems required for operation.

Initial process integration and characterization was done to determine a proper fabrication process; however, the yield of successful test structures was low. Problems arose in the first round of fabrication because the protective parylene was etched through so that during the final release step, the silicon etch that was supposed to be protected was instead etched away. The photolithography mask was revised in an attempt to prevent this occurrence with other wafers. Further hypothesized problems include keyholes forming when the top of the trench is closed before the trench is filled during the parylene deposition in the deep trenches of the leg link structure. The wafers from the first fabrication run have been analyzed and more wafers are being processed to duplicate or improve upon the results found.

2. Experiment

Test structures from the first fabrication run performed by Dr. Kenn Oldham were diced and analyzed using scanning electron microscopy (SEM). To ensure the charging of the parylene would not affect the quality of the SEM images, a thin coating of sputtered gold was added to each test structure. Trenches of various widths that had been filled with parylene were of particular interest to determine if keyholes had formed during deposition. In addition to the SEM, these test structures were photographed using the optical microscope to look for defects and flaws in the parylene coating, especially at the intersection junction of critical features.

In order to perform testing of the parylene fabrication process, prototype test actuators were made from silicon dioxide, platinum-coated silicon dioxide, or a lead zirconate titanate (PZT) stack; all approximately 5000 Å thick. These structures are patterned to represent the size and shape of piezoelectric actuators used to create large in-plane rotational joints. The silicon dioxide test actuators were deposited using a plasma-enhanced chemical vapor deposition (PECVD) process to achieve a depth of approximately 5000 Å, and then patterned and etched via reactive ion etching (RIE). The platinum-coated silicon dioxide test actuators were formed by first depositing 5000 Å of silicon dioxide via PECVD and then sputtering a 200 Å titanium adhesion layer followed by an 820 Å platinum layer. This was then patterned and etched using ion milling and RIE. The PZT layers were deposited as described in section 1, and then patterned and etched using ion milling and RIE.

After deposition and subsequent patterning of these test actuators, narrow trenches were etched into the silicon wafer with deep RIE (DRIE). The etch depth was one of the critical features to assess during this characterization and was varied from 75 to 300 μm. To characterize the trench etch depth, a Tencor

Stylus profilometer was used to measure wide features (~1 50 μm) to give an assessment of the etch depth in the narrow trenches. The narrow features can be measured with conventional metrology tools such as optical and stylus profilometry. Furthermore, RIE lag severely degrades the etch rate in high aspect ratio narrow trenches; therefore, measuring the wider features only gives an estimate of the etch depth. Accurate measurements are performed using cross-section SEM measurements after fabrication.

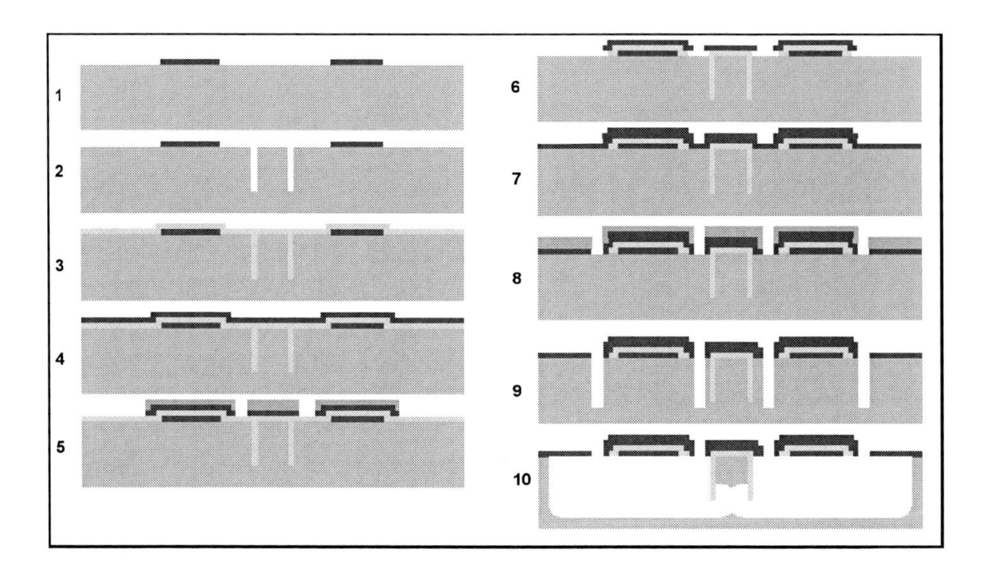

Figure 2. Parylene fabrication process: 1. Pattern (dummy) actuators; 2. Etch deep trenches; 3. Coat parylene; 4. Deposit oxide hard mask; 5. Pattern photoresist for parylene etch; 6. Etch parylene and photoresist; 7. Deposit additional oxide; 8. Pattern oxide, thin photoresist; 9. Etch deep trenches; and 10. XeF_2 release.

Following the silicon DRIE, the wafer is then coated with hexamethyldisilazane (HMDS) and loaded into a parylene deposition system. The parylene is deposited to a thickness of approximately 7 μm in order to fill the trenches. After the parylene coating, a 100 nm thick silicon dioxide is deposited with a low-temperature PECVD at 160 °C. The silicon dioxide is then patterned and etched using RIE with a fluorinated plasma, and serves as a hard mask for the removal of parylene via an oxygen plasma etch. After the exposed parylene has been removed, a second layer of PECVD silicon dioxide is deposited at 160 °C, and subsequently patterned and etched via RIE to expose the base silicon. This silicon dioxide functions as a hard mask in order to prevent exposing the parylene to further oxygen plasma during photoresist removal. Trenches are then etched into the silicon via DRIE approximately 10 μm deeper than the first DRIE to allow for the parylene-filled trenches to be undercut. This undercut is achieved by exposing the wafer to xenon diflouride (XeF_2) gas, which isotropically etches silicon that is not protected with parylene. Finally, a short RIE is done to remove the remaining oxide on the surface. The process flow is shown graphically in figure 2.

A labeled sample actuator/flexure/leg-link test structure is shown in figure 3 [2]. The structures pictured in this report also include a D-Flexure, which is an added connection between the anchor and the leg-link (parallel to the flexure) that provides torsional stiffness.

These leg links are also combined in a series configuration in order to provide the large range of angular motion. This combined angular motion of approximately 45° of rotation at the joint mimics insect-like mobility.

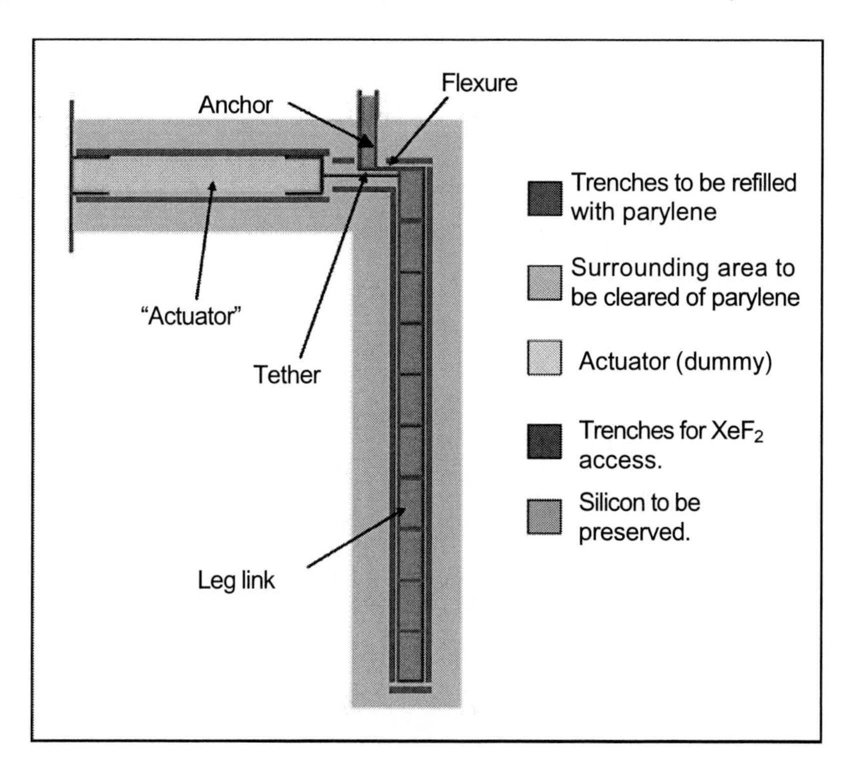

Figure 3. The sample actuator/flexure/leg-link test structure. In figures 4–8 the structures also contain a D-Flexure in parallel to the flexure.

3. Results and Discussion

Results from the initial testing of the process fall into four categories: trench junction failures, parylene adhesion and etching, trench analysis, and photolithography. These categories are discussed in depth below.

3.1. Trench Junctions

Observations of the first processed wafers indicate that over half of the structures exhibited defects at junctions where four trenches meet as seen in photographs from an optical microscope in figures 4 and 5.

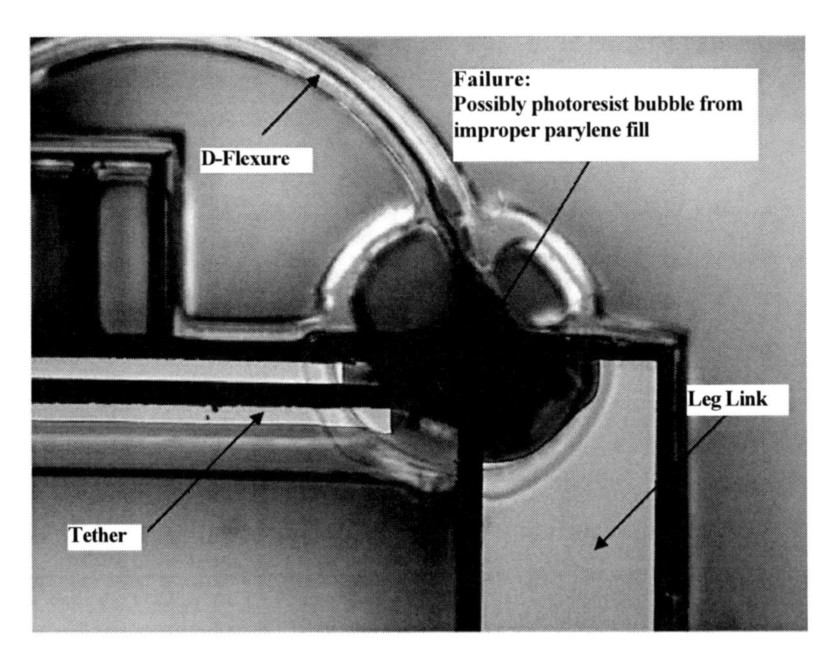

Figure 4. This photograph from the optical microscope illustrates failure at the junction between the flexure and the leg link, possibly from improper parylene fill or photoresist bubbles.

The defects are of specific concern because these failures are occurring at connection points of the flexure, which are essential in providing a connection to the anchor, as well as creating torsional rigidity and in-plane flexibility. If this connection fails, then the leg link could be completely separated from the actuator or twist out of plane.

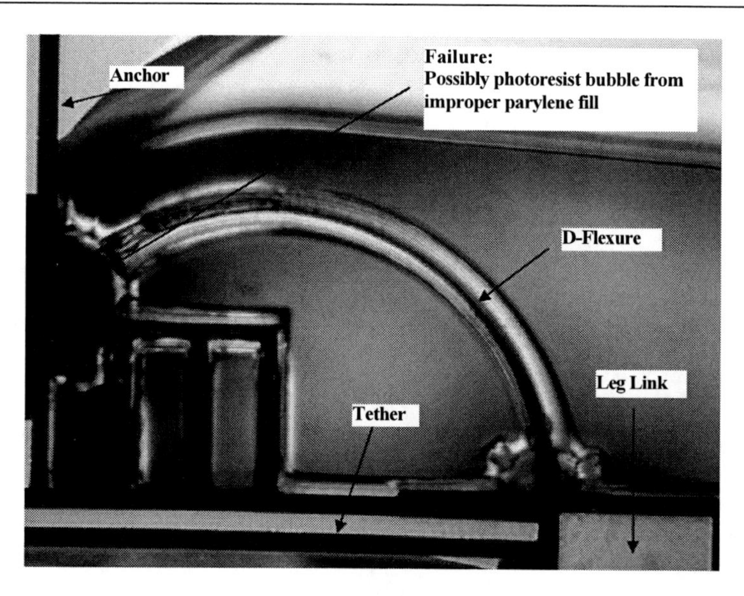

Figure 5. This photograph from the optical microscope illustrates a failure at the junction between the D-flexure and the anchor, possibly from improper parylene fill or photoresist bubbles.

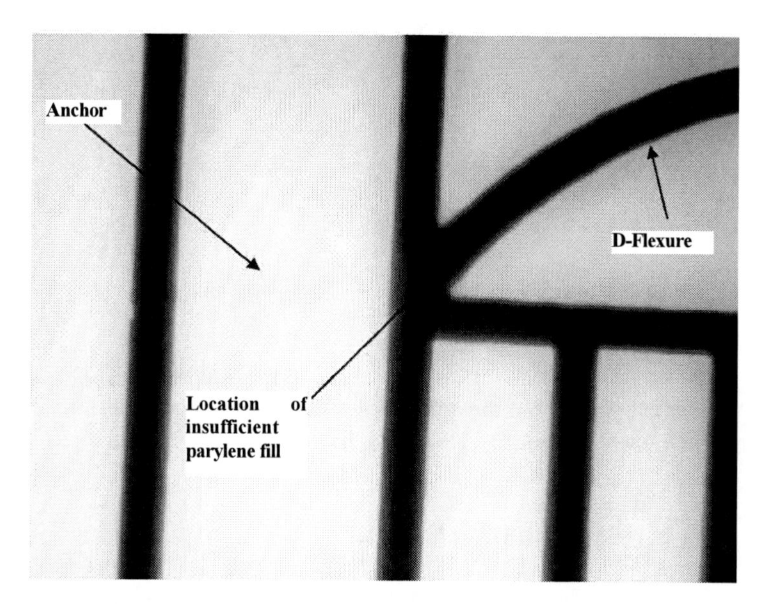

Figure 6. This photograph from the optical microscope shows insufficient parylene fill in the dark black areas near the four-trench junction at the anchor and D-Flexure intersection.

Process or design changes are necessary to correct these issues. Options include depositing thicker parylene coatings to properly fill the junctions, decreasing the width of trenches near junctions, and avoiding four- trench junctions where possible.

Insight into the failure at the four-trench junctions may be seen immediately after parylene coating in figures 6 and 7, showing that the parylene did not fully coat the entire junction. These images indicate that a thicker parylene coating should be deposited to properly fill the entire trench at the junction intersections or a mask change should be made to decrease the width of the trenches near the intersections.

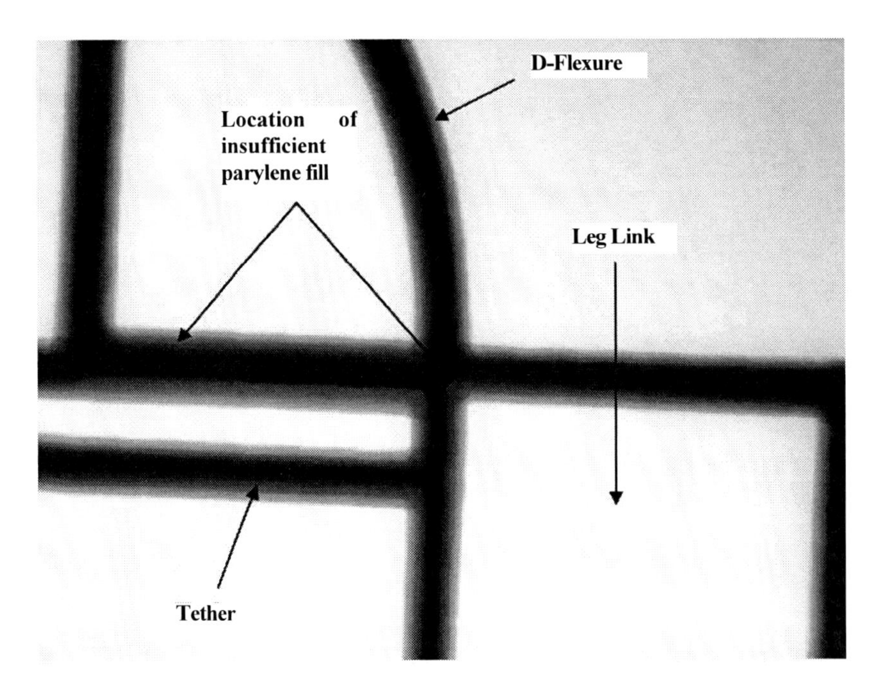

Figure 7. This photograph from the optical microscope shows insufficient parylene fill in the dark black areas near the D-Flexure-leg link four-trench junction.

3.2. Parylene Adhesion and Etching

Parylene adhesion to platinum and PZT was of initial concern as proper adhesion is critical to the success of the actuator joints. Parylene adhesion to platinum was tested utilizing a laminated composite consisting of silicon dioxide coated with a thin titanium layer and 85 nm of platinum. After patterning of the test actuators, parylene was deposited as with the silicon dioxide test actuators. The initial parylene coating on platinum proved successful with a uniform coating having the exact same appearance and characteristics as the coatings on silicon dioxide. In addition, a parylene coating was applied to a composite stack of SiO_2/Ti/Pt/PZT/Pt with thicknesses similar to the actuators used in reference 1. Similar to the previous results, the parylene coating was uniform without defects or signs of delamination.

Parylene etching has presented a problem in this process because the original methods of using Axcellis Downstream Asher resulted in more than 9 μm of parylene undercut due to the isotropic nature of the etching process. This undercut is more than 3 μm larger than the tolerable undercut in the design. For this reason, the etching process was changed to use an anisotropic RIE in order to prevent this excessive undercut [3]. Using a Lam 590 parallel plate RIE system, the process showed considerable improvement with less than a 2 μm undercut. However, the process leaves behind residues on the exposed silicon surface, as seen in figure 8.

At this time, it is unknown whether or not this residue will pose a problem with the final process steps. Further analysis needs to be done to determine if this is a significant issue.

Figure 8. Surface residue on exposed silicon after anisotropic oxygen plasma RIE Lam 590.

After patterning the parylene, samples showed no immediate parylene delamination; however, after an hour or more, several samples began to delaminate. This delamination was seen on wafers with silicon oxide dummy actuators as well as wafers with PZT dummy actuators, so it is unclear if the dummy actuator material contributes at all to this delamination. Further investigation is needed in order to determine if the parylene coating and patterning process is time sensitive. Furthermore, since the parylene coating was done outside of the cleanroom, the delamination could also be attributed to the unclean environment or to the time between preparing the wafer surface with HMDS and actually depositing the parylene. The entire wafer was not delaminated and some of the test structures on the PZT coated wafer have been successfully released.

3.3. Trench Analysis

Proximity of trenches as well as trench width has an effect on trench depth and etch profile for silicon DRIEs performed with the Bosch process [4, 5, 6]. In order to characterize this effect for our specific purposes, trenches spaced 420 μm apart (figure 9) and 76 μm apart (figure 10) were run through the same DRIE and were then filled with parylene, diced, and compared using the SEM.

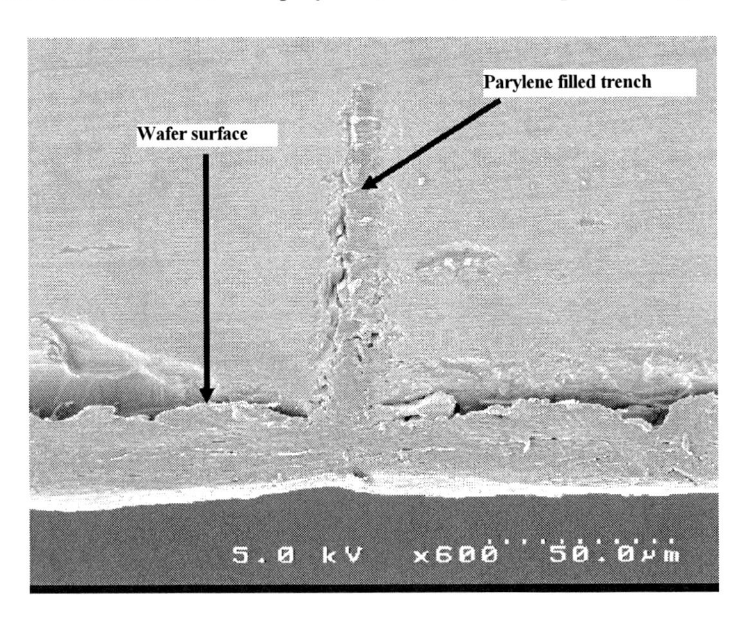

Figure 9. This photograph from the SEM shows a parylene filled trench 5 μm wide. The etch depth here is approximately 82 μm.

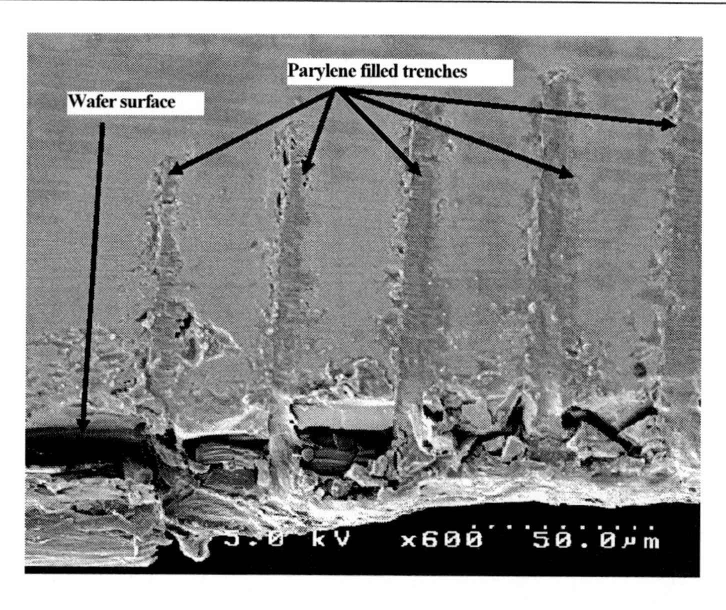

Figure 10. This photograph from the SEM shows the parylene filled trenches of widths of 3, 4, 5, 6, and 7 µm spaced 30 µm apart. The etch depth is approximately 76 µm for the 5 µm wide trench.

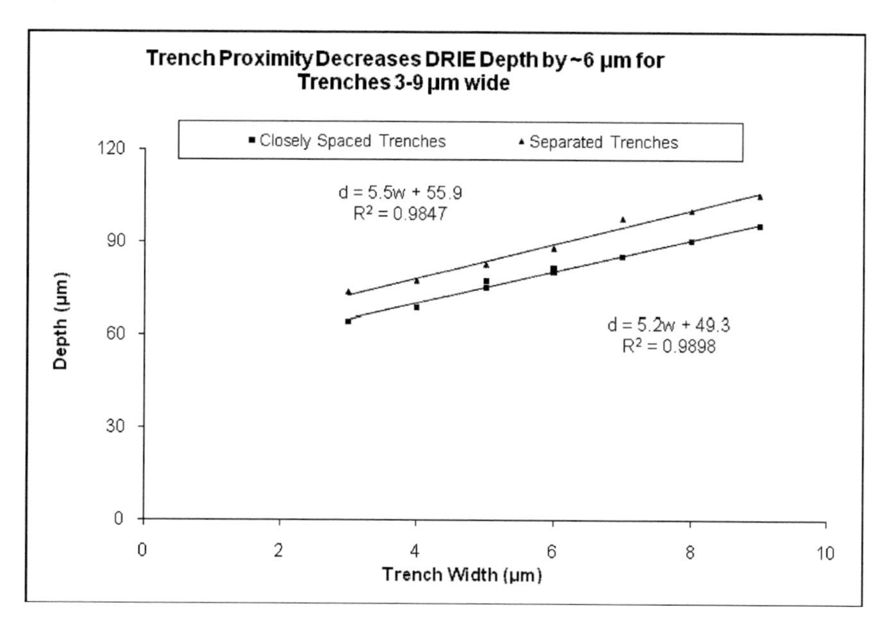

Figure 11. The graph shows that trenches that are close together result in more shallow etch profiles and that trench depth is linearly dependent on trench width within the range of etch.

Measurements of these trenches show that for trenches within the range of 3–9 μm wide and 60–1 20 μm deep, trench depth is approximately linearly dependent on trench width and also offset by trench proximity by approximately 6 μm as seen in figure 11.

Although the affect of trench proximity is small, it is still something that should be considered during design. It is especially important when considering placing trenches near large openings, which can locally deplete the etching species [4].

3.4. Photolithography

Several processing challenges occurred that were related to photolithography and will be described in this section. These challenges, though their effects were unfavorable, should be easily accounted for in subsequent processing.

During the last photolithography step, after the etching of parylene, three wafers were placed in acetone to remove photoresist due to insufficient coverage during the spinning of photoresist on the wafers. It appears that on these wafers, the acetone caused delamination at the etch fronts of the parylene-coated layer. To prevent delamination in future processing, the last photolithography step must be completed correctly on the first attempt as the photoresist cannot be removed without etching or delaminating the parylene coating. If photoresist must be removed, PRS3 000 may work, but this has not been proven successful and it would be best if it could be avoided. A manual spin may be the best option for coating the wafer in this photolithography step in order to ensure good coverage over all of the topography on the wafer.

Finally, during the second DRIE, one of the wafers had been overetched. The photoresist was completely removed during the DRIE and the remaining parylene was removed during the DRIE and afterwards when the photoresist was supposed to be removed using a short oxygen plasma. In the original process flow, the photoresist was supposed to be thinned prior to the second DRIE; however, this step is not needed and can result in overetching of the structures.

4. Conclusions

From observations of previously processed wafers as well as wafers being currently processed, it is apparent that changes to the process are needed. First, the photoresist thinning step prior to the second deep silicon etch should be removed in order to prevent overetching. Second, the last photolithography step should be manually spun in order to ensure full coverage and avoid exposing the parylene to acetone or oxygen plasma, which can etch and cause delamination of the parylene. Third, in order to etch the parylene, an anisotropic RIE should be used to prevent undercut, followed by a downstream oxygen plasma etch to prevent the surface residue (if it proves to be problematic). Fourth, a thicker parylene coating should be deposited to successfully cover the four-trench junctions and prevent photoresist bubbles and trench junction failures. Finally, further work needs to be done to determine the cause for the delayed parylene delamination observed.

There are also design changes that may be made, such as eliminating four-trench junctions where possible. This mask change may be aided by the fact that the electrical interconnects are still being tweaked to prevent adding excessive stiffness.

Future work will also include duplicating the full processing on wafers with oxide dummy structures and determining yield, and eventually, conducting functional device testing.

Acronyms

ARL	U.S. Army Research Laboratory
DRIE	deep reactive-ion etching
HMDS	Hexamethyldisilazane
PECVD	plasma enhanced chemical vapor deposition
piezoMEMS	piezo-microelectromechanical systems
PZT	lead zirconate titanate
RIE	reactive ion etching
SEM	scanning electron microscopy
XeF_2	xenon difloride

Acknowledgments

The authors would like to acknowledge Richard Piekarz, Joel Martin, and Eugene Zakar of the Sensors and Electron Devices Directorate, U.S. Army Research Laboratory and Brian Power of General Technical Services.

References

[1] Oldham, K.; Pulskamp, J.; Polcawich, R. G.; and Dubey, M. Thin-Film PZT Lateral Actuators with Extended Stroke. *J. MEMS* 2008, 17, 890–899.

[2] Oldham, K. *Parylene Mechanisms for Integration with Piezoelectric Thin-Film Actuators*; ORISE final Report, 2008.

[3] Meng, E.; Li, P.; Tai, Y. Plasma removal of Parylene C. *Micromechanics and Microengineering* 2008, 18.

[4] Ayón, A. A.; Braff, R.; Lin, C. C.; Sawin, H. H.; Schmidt, M. A. Characterization of a Time Multiplexed Inductively Coupled Plasma Etcher. *Journal of The Electrochemical Society* 1999, 146 (1), 339–349.

[5] Ayón, A.A.; Zhang, X.; Khanna, R. Anisotropic silicon trenches 300-500μm deep employing time multiplexed deep etching (TMDE). *Sensors and Actuators* 2001, 91, 381–385.

[6] Blauw, M. A.; Zijlstra, T.; van der Drifta, E. Balancing the etching and passivation in time- multiplexed deep dry etching of silicon. *J. Vac. Sci. Tech. B* 2001, 19 (6).

In: New Robotics Research ISBN 978-1-60741-093-5
Editors: E.D. Wagner et al, pp. 77-106

Chapter 5

REAL TIME PANORAMIC DEPTH IMAGING FROM MULTIPERSPECTIVE PANORAMAS USING STANDARD CAMERAS

Peter Peer* and Franc Solina
University of Ljubljana
Faculty of Computer and Information Science
Tržaška 25, SI-1000 Ljubljana, Slovenia

Abstract

Recently we have presented a system for panoramic depth imaging with a single standard camera. The system is mosaic-based, which means that we use a single standard rotating camera and assemble the captured images in a multiperspective panoramic image. Due to a setoff of the camera's optical center from the rotational center of the system we are able to capture the motion parallax effect from a single sweep around the rotational center, which enables the stereo reconstruction. One of the problems of such a system is the fact that we cannot generate a stereo pair of images in real time. This chapter presents a possible solution to this problem, which is based on simultaneously using many standard cameras. We perform simulations on real scene images to establish the quality of new sensor results in comparison to results obtained with the old sensor. The goal of the chapter is to reveal whether the new sensor can be

*E-mail address: peter.peer@fri.uni-lj.si. Tel. +386 1 4768 878, Fax +386 1 4264 647. (Corresponding author)

used for real time capturing of panoramic depth images and consequently for autonomous navigation of a mobile robot in a room. In particular, we focus on the real time generation of panoramic stereo pairs since the calculation of depth images can already be run in real time. The basic panoramic depth imaging system and its real time extension are comprehensively analysed and compared. The analyses reveal a number of interesting properties of the systems. According to the basic system accuracy we definitely can use the system for autonomous robot localization and navigation tasks. The assumptions made in the real time extension of the basic system are proved to be correct, but the accuracy of the new sensor generally deteriorates in comparison to the basic sensor.

Keywords: Computer vision, Stereo vision, Reconstruction, Depth image, Multiperspective panoramic image, Mosaicing, Motion parallax effect, Standard camera, Real time, Depth sensor

1. Introduction

Real time panoramic depth imaging from multiperspective panoramas is an issue that is not well covered in the literature. There have been discussions [11] about it, but nothing has been done in practice so far, at least not by using the idea of multiperspective panoramas (the definition is given in the beginning of the next section).

Generally, mosaic-based procedures for building panoramic images [1, 3, 5, 11, 12, 17, 18, 21] can be marked as non-central, they do not execute in real time and they give high resolution results. Thus mosaicing is not appropriate for capturing dynamic scenes. The main advantage of these procedures over other panoramic imaging systems (like catadioptric systems [25]) is the ability to generate high resolution results. But high resolution results are essential for effective depth recovery based on the stereo effect.

In this chapter we focus on real time generation of a multiperspective panoramic stereo pair. Calculation of depth images from such stereo pairs already runs in real time [24].

In [12] we presented a system for panoramic depth imaging with a single standard camera. We reported a number of interesting properties of the system. In Sec. 3. we summarize them and report a number of new properties and experimental results [13, 14], all in order that the reader gets a good picture about the system. Our main motivation was to establish the quality of results for autonomous robot localization and navigation tasks. In this chapter we go a step further and evaluate a real time extension of the system. To do that, we first in Sec. 4. discuss how the time needed for the generation of a symmetric panoramic stereo pair can be dramatically reduced. In Sec. 5. we go even further and explain how we can achieve real time execution. Sec. 6. presents the depth reconstruction equation for the new setup. The epipolar constraint is discussed in Sec. 7. The evaluation of results is given in Sec. 8. We end the chapter with conclusions in Sec. 9. But let us first start by saying a few words about related work.

2. Related Work

One way to build panoramic images is by taking one pixel column out of a captured image and mosaicing the columns. Such panoramic images are called multiperspective panoramic images [21]. The crucial property of two or more multiperspective panoramic images is that they capture the information about the motion parallax effect, since the columns forming the panoramic images are captured from different perspectives.

However, multiperspective panoramic images are not something new to the vision community [21]: they are a special case of *multiperspective panoramic images for cel animation* [28], a special case of *crossed-slits (X-slits) projection* [2, 7, 29], they are very similar to images generated by a procedure called *multiple-center-of-projection* [19], by the *manifold projection* procedure [17] and by the *circular projection* procedure [15, 16]. The principle of constructing multiperspective panoramic images is also very similar to the *linear pushbroom camera* principle for creating panoramic images [8].

The papers [11, 21, 22], which are closest to our work about the panoramic depth imaging with a single standard camera [12, 13, 14] seem to lack two things: a comprehensive analysis of 1) the system's capabilities and 2) the corresponding points search using the epipolar constraint. Therefore, the focus of the next section is on these two issues. While in [11] the authors searched for corresponding points by tracking the feature from the column building the first panorama to the column building the second panorama, the authors in [21] used an upgraded *plane sweep stereo* procedure. A key idea behind the approach in [22] is that it enables optimizing the input to traditional computer vision algorithms for searching the correspondences in order to produce superior results.

There are of course other somehow similar possibilities to acquire panoramic depth images. Let us briefly present the system called SOS (Stereo Omnidirectional System) [20, 26, 27]. The SOS system uses standard cameras and produces panoramic depth images in real time, but here all the similarities with our system end. The system consists of 20 stereo units and each unit consists of 3 standard cameras. The units are arranged on each plane of a regular icosahedron. The basic principle for generating panoramic depth images is as follows: Each unit captures three images from which a normal (not panoramic) depth image is computed on a computer with more CPU units. Then the generated depth images are registered into the panoramic depth image. Obviously, the authors use a different concept than we do: they first build standard depth images and then the panoramic depth image, while we first build panoramic images and then the panoramic depth image. So, from this point on, we are interested only in panoramic depth imaging from multiperspective panoramas.

In order to capture stereo panoramic images of dynamic scenes Peleg et al. [18] presented a theory for construction of a special mirror and lens such that viewing the scene through this mirror or lens creates the same rays as those used with the rotating cameras. To our knowledge the systems are still not constructed, while, according to authors, a lot of practical issues need to be solved before a camera is built.

As mentioned, the real time extension of our system to our knowledge has not been suggested and evaluated before.

3. About the Basic System

Our basic system for generating panoramic depth images with a single standard camera captures a stereo pair of images while the camera rotates around the center of the system in a horizontal plane. The motion parallax effect which enables the reconstruction can be captured because of the offset of the cameras' optical center from the systems' rotational center (see the small photograph within Fig. 1). The camera is moving around the rotational center in angular steps corresponding to one vertical pixel column of the captured standard image. A symmetric pair of panoramic stereo images are generated so that one column on the right side of the captured image contributes to the left eye panoramic image and the symmetric column on the left side of the captured image contributes to the right eye panoramic image. So, we are building each panoramic image from just a single pixel column of the captured image. Thus, we get a symmetric pair of stereo panoramic images, which yields a reconstruction with optimal characteristics (simple image row epipolar constraint [9, 10, 21, 23] and minimal reconstruction error [22]).

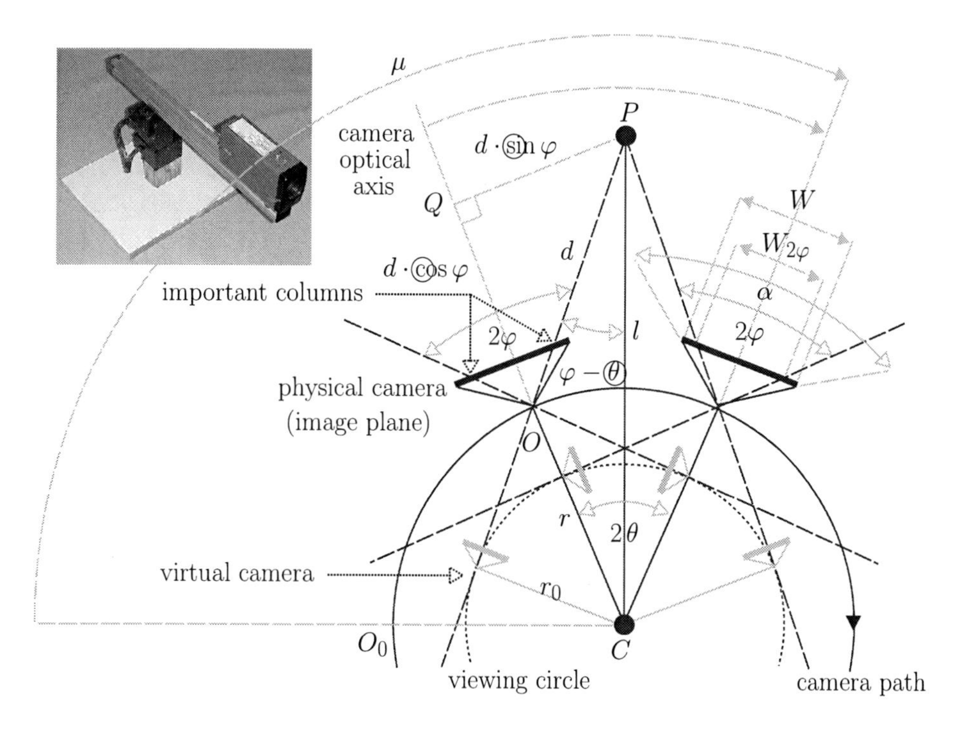

Figure 1. Geometry of our basic system for constructing multiperspective panoramic images. Note that a ground-plan is presented; the viewing circle extends in 3D to the viewing cylinder. The optical axis of the camera is kept horizontal. In the small photograph the hardware part of the system is shown.

The geometry of our basic system for creating multiperspective panoramic images is shown in Fig. 1. The panoramic images are then used as the input to create panoramic depth images. Point C denotes the system's rotational center around which the camera is rotated. The offset of the camera's optical center from the rotational center C is denoted as r, describing the radius of the circular path of the camera. The camera is looking outward from the rotational center. The optical center of the camera is marked with O. The column of pixels that is sewn in the panoramic image contains the projection of point P on the scene. The distance from point P to point C is the depth l, while the distance from point P to point O is denoted by d. Further, θ is the angle between the line defined by points C and O and the line defined by points C and P. In the panoramic image the horizontal axis represents the path of the camera. The axis is spanned by μ and defined by point C, a starting point O_0, where we start capturing the panoramic image, and the current point O. φ denotes the angle between the line defined by point O and the middle column of pixels of the image captured by the physical camera looking outward from the rotational center (the latter column contains the projection of the point Q), and the line defined by point O and the column of pixels that will be mosaiced into the panoramic image (the latter column contains the projection of the point P). Angle φ can be thought of as a reduction of the camera's horizontal view angle α.

The main conclusions made throughout the analysis are (see [12, 13, 14] for discussion about each item):

- The geometry of capturing multiperspective panoramic images can be described with a pair of parameters (r, φ). By increasing (decreasing) each of them, we increase (decrease) the baseline $(2r_0)$ of our stereo system.
- The stereo pair acquisition procedure with only one standard camera cannot be executed in real time.
- The epipolar constraint in case of symmetric stereo pair of panoramic images, which we use in the reconstruction process, is very simple: epipolar lines are image rows.
- The parameters of the system should be estimated as precisely as possible, since already a small difference can cause a big difference in the reconstruction accuracy of the system.
- We can effectively constrain the search space on the epipolar line. This follows directly from the interpretation of the equation for depth estimation l, while other rules for constraining the search space, known from traditional stereo vision systems, can also be applied in addition to the basic constraint. An example of such rule is to seek for the neighboring pair of corresponding points only from the previously found correspondence on.
- The confidence in the estimated depth is variable: 1) the bigger the slope of the function l, the smaller the confidence in the estimated depth (one-

pixel error[1] Δl gets bigger) and 2) the bigger the value φ for each camera (α), the bigger the number of possible depth estimates and consequently the bigger the confidence.

- We can influence the parameter θ_0[2] by varying the resolution of captured images or by varying the horizontal view angle α.
- By varying the radius r, we vary the biggest possible and sensible depth estimation l and the size of the one-pixel error Δl.
- The bigger the value α, the smaller the horizontal resolution of panoramic images at fixed resolution of captured images. Consequently, the number of possible depth estimates per one degree gets lower.
- In practice, from the autonomous robot localization and navigation system point of view, we should define the upper boundary of the allowed one-pixel error size Δl.
- The contribution of the vertical reconstruction[3] is small in general, but has a positive influence on the overall results.
- The numbers of possible depth estimates are very similar for different cameras (α) at fixed resolution of the captured images.
- The size of the one-pixel error Δl is also similar at similar number of possible depth estimates for different cameras.
- The reconstruction process (after the stereo pair has been generated) can be executed in real time.
- The reconstructed points lie on concentric circles centered in the center of rotation and the distance between circles (the one-pixel error Δl) increases the further away they lie from the center.

[1] As the images are discrete, we like to know what is the value of the error in the depth estimation if we miss the right corresponding point for one pixel.

[2] Our system works by moving the camera for the angle corresponding to one pixel column of the captured image.

[3] In contrast to [12], we have incorporated the vertical view angle β into the equation for depth estimation l in [13].

- The linear model for estimation of angle φ have been proved better for a given set of parameters in comparison to the non-linear model[4].
- We can achieve similar reconstruction accuracy with panoramas build from only one-pixel column of the captured images in different rooms, even with different cameras.
- The remaining error in accuracy could be attributed to a number of possible reasons (e.g. to the fact that we are limited with the number of possible depth estimates, which are approximations of the real distances etc.).
- Processing undistorted images in general brings better though comparable results, but undistorting the sequence can be time expensive task and we are forced to re-estimate some parameters of the system after the distortion is corrected.
- According to the basic system accuracy, we definitely can use it in autonomous robot localization and navigation tasks.

This system however cannot generate panoramic stereo pair in real time. To illustrate this fact, we can write down the following example from practice: if the system builds a panoramic stereo pair from standard images with resolution of 160×120 pixels, using a camera with the horizontal view angle $\alpha = 34°$, it needs around 15 minutes to complete the task.

The first idea about how to capture a stereo pair quicker is to generate panoramic images from wider vertical stripes instead of just one column.

4. Building Panoramic Images from Wider Stripes

This task is by all means much faster, but at the same time we have to make a compromise between the speed of the capturing task and the quality of the stereo pair. First of all, the wider the stripes are, the more obvious are the stitches between the stripes in the panoramic image. Then, with the wider stripe the difference between the amount of lens distortion of the first and the last pixel columns in the stripe is more noticeable. But the real problem arises from the fact that stripes introduce a property, which influences the coverage of the scene,

[4]The non-linear model contains the focal lenght f explicitly.

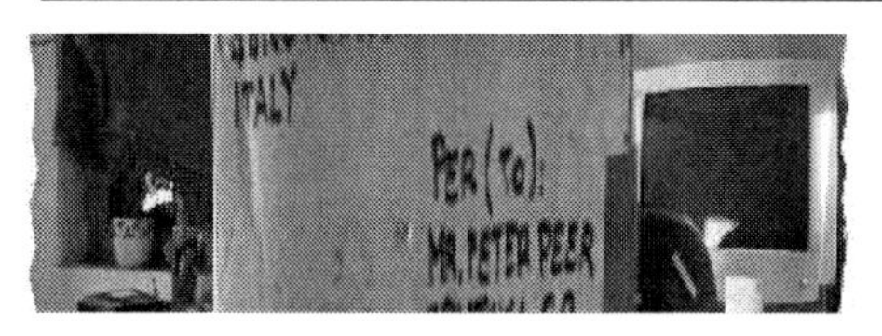

Figure 2. The wider is the stripe, the more scene points are not captured in the panoramic image: the left panoramic image was built from single columns, while the right panoramic image was built from 14 columns wide stripes. Note how very distant scene points are well captured in both examples and how some nearby scene points (text on the box) are not captured in the second example.

as presented in the next section. As we show in the experimental results (Sec. 8.), we are satisfied with the result when we use 14 pixel columns wide stripes and we think that it represents a good compromise. This statement is naturally highly related to the camera that we use.

In that case, the horizontal view angle of the camera is 34°, where 14 columns represent the angle of approximately 3°, the building process takes approximately 14 times less, i.e. around a minute.

4.1. Property of Using Stripes

If we observe the panoramic image built from stripes closely, we can notice that the image is not perfect. In this case we are not referring to stitches nor lens distortion. These problems are present, but are not too disturbing. Another problem can be noticed on close objects on the scene, which have a nice texture on them (like text). In such a case we can see that some points on the scene are not captured (Fig. 2). Of course, this is partly because of the fact that we are dealing with the images, which are discrete (for example: we cannot take a half of a pixel), but if we take a look at the geometry of the system (Fig. 1), we can see that this is not the only reason. If we consider two successive steps of the system, we can see that the stripes that contribute to the panoramic image do not cover all the scene.

The drawing in Fig. 3, shows the formation of the panoramic image with respect to the light rays (compare with Fig. 1). The drawing is done for the actual case of horizontal view angle $\alpha = 34°$ to illustrate the problem of uncaptured scene points. The large circle represents the camera path. The small circle represents the viewing circle. Point C in the middle of the circles represents the center of rotation. Point O represents the position of the optical center on the camera path. A middle line going from the optical center O outwards represents the light ray that is in the middle of light rays that form a stripe in each captured image that contributes to the panoramic image. A line tangent to the viewing circle is the extension of that light ray towards the virtual camera. Two lines on each side of this light ray going from the optical center O outwards represent bordering light rays that form a stripe in each captured image that contributes to the panoramic image.

Let us first examine the example given on the right side in Fig. 2, where the shift angle corresponds to the stripe of 14 pixel columns of the captured image (W_s=14): $\alpha = 34°$, $r = 30$ cm, $2\varphi_{\min} = 27.082501°$, $2\varphi_{\max} = 32.842499°$ (these two values $\varphi_{\min}$ and $\varphi_{\max}$ define the stripe, which consists of more pixel columns, similarly as φ defined one pixel column; for more information see Sec. 6.), $\theta_0 = W_s \cdot 0.205714°$. The drawing is presented in Fig. 3. Since each stripe of the captured image is formed from more light rays, which also circumscribe a similar angle to the systems' shift angle (compare definitions of $\theta_0(\alpha)$ and $\theta_0(\varepsilon)$ given below), the middle light ray and the border light rays are presented for each stripe that contributes to the panoramic image. We can observe two properties on this drawing (note the zoomed in detail in it): 1) We really cannot capture all the scene points, since there is a gap between two successive stripes that contribute to the panoramic image. 2) With the growing distance from the rotational center the bordering light rays of two successive stripes are getting nearer until they intersect. Generally, the shift angle and the angle corresponding to the stripe of the captured image should be the same, but they are not due to the limited accuracy of our rotational arm. In this general and ideal case the bordering light rays of two successive stripes would be parallel (for example, imagine that both angles are 90°). Since we deal here with two independent discretizations, namely with the discretization of the standard image (pixels; $\theta_0(\alpha)$) on one side and with the discretization of the rotational arm (the minimal step of the rotational arm defined by its maximal accuracy; $\theta_0(\varepsilon)$) on the other side, it is almost impossible to achieve the ideal case ($\theta_0(\alpha) = \theta_0(\varepsilon)$). Because of these discretizations the bordering light rays of two successive stripes are normally getting nearer ($\theta_0(\alpha) > \theta_0(\varepsilon)$) or further apart ($\theta_0(\alpha) < \theta_0(\varepsilon)$).

Already the first property by itself reveals that there is always a part of the scene which cannot be captured. And since the effect loses on significance with a bigger distance from the rotational center, the distant points on the scene look well captured (Fig. 2), also since the bigger distance to the object results in lower resolution of the object. On the other hand, in our case, the bordering light rays of two successive stripes are getting nearer with bigger distance, which means that this effect is getting even smaller with increased distance. This also means that in the far distance the bordering light rays are intersecting, but since the intersection point is far away the effect of repeatability of scene points in the panoramic image is not noticeable, again because with the bigger distance of the system from the object its resolution in the image gets lower.

A similar effect is present when we build panoramic images from only one pixel column of each captured image, while one pixel width is not infinitesimal, although the part of the panoramic image presented on the left side in Fig. 2 looks faultless. In this case the gaps are much smaller.

So, we can write one more conclusion with respect to this property. Namely, the wider is the stripe, the more scene points are not captured in the panoramic image (Figs. 2 and 3). And this holds regardless of the position of the stripe in the captured image. We can take a stripe from the middle of the captured image or from the edge of it, but the presented property would still be seen on the resulting panoramic image. Naturally, by using single columns we achieve best possible result (Fig. 2), though still not perfect, since the described property still holds, but is not so obvious. By using smaller r the gaps are also smaller, which means that with smaller r we cover more scene points.

Another conclusion is that if we want to capture the parallax effect, we have to accept the fact that not all the scene points are captured. When it comes to scene reconstruction, we can address the missing scene information problem in a similar manner as in the case of a normal, non-panoramic stereo system (e.g. problem of occlusions).

5. Achieving Real Time

If we use stripes instead of single columns from each captured image, we can drastically reduce the time needed to build a panoramic image. But we still cannot build a panoramic image in real time.

The idea for a real time panoramic sensor is actually very simple. In our old system the panoramic image is build by means of moving the standard camera for a very small angle along a predefined circular path. If we could have a camera on each position on the circular path, we could build the panoramic image in real time. But unfortunately in practice we cannot put so many cameras so close together (with respect to a reasonable size of radius r). If we build a panoramic

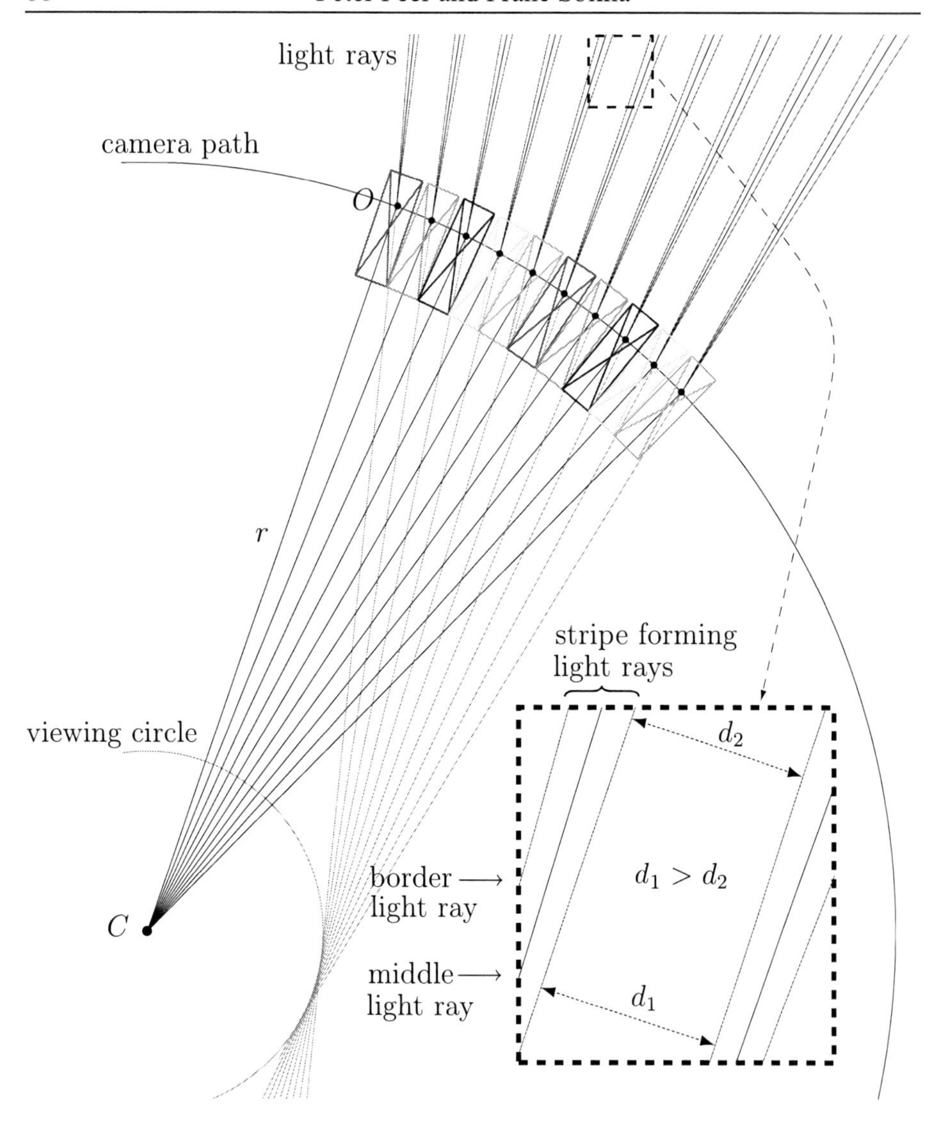

Figure 3. The drawing shows the formation of the panoramic image with respect to the light rays. In this case the shift angle corresponds to 14 pixel columns of the captured image. For detailed description see Secs. 4.1. and 5..

image from captured images with the resolution of 160×120 pixels, then we have to put the cameras with the horizontal view angle $\alpha = 34°$ 0.205714°

apart from each other and we need 360/0.205714≐1750 cameras.

In the case when we use stripes, the presented numbers get more reasonable. A 14 column stripe suggests that the cameras would be 2.879999° apart from each other and we would need 125 cameras to cover the whole circular path. If we use a camera with a wider horizontal view angle (e.g. $\alpha = 90°$), we need less cameras (e.g. 46). The new sensor does not need any moving parts, which means that we do not deal with mechanical vibrations nor are we limited with the radius of the circle on which the cameras are fixed. The last statement about the radius enables us to make the sensor out of standard cameras that are available on the market.

Fig. 3 also shows the drawing of a real time sensor. The boxes with crosses represent the physical cameras positioned with respect to the center of rotation. The sensor is made out of cameras presented in the photograph within Fig. 1 ($\alpha = 34°$), where the angle between successive cameras corresponds to 14 columns of the captured image with the resolution of 160×120 pixels. This camera is probably much to big to be used in a real sensor (the ground plan dimensions of the camera are 14.5×4.5 cm), since r in this case should be set to at least 100 cm. On the other hand, we could set r to 30 cm and this implies that the camera should be smaller for $3.\overline{3}$ times (in this case the ground plan dimensions of the camera should be 4.35×1.35 cm). We have simulated such sensor with our camera in order to determine its accuracy. Namely, nowadays there are many practical solutions how to increase processing power, but before we build a real system, we have to know whether the accuracy of the proposed system is satisfactory.

The experimental results obtained by rotating a single camera are given in Sec. 8., where we have also tested other cameras, different widths of the stripes etc. If we built the sensor, we would normally use board cameras, where the lens, as the biggest part, is even smaller and the board to which the lens is attached is flexible, so that we are able to put the lenses completely together.

6. Stereo Reconstruction from Stripes

By following the sinus law for triangles, we can simply write the equation for the depth l as (Fig. 1):

$$l = \frac{r \cdot \sin\varphi}{\sin(\varphi - \theta)}. \tag{1}$$

This equation holds if we do the reconstruction based on the symmetric pair of stereo panoramic images built from one pixel columns of the captured image.

But when we use stripes, we have to adopt the equations according to the new building process. In this case we take symmetric stripes instead of symmetric columns from the captured image. While the column was defined by the angle φ, the stripe is defined by two such angles: $\varphi_{\min}$ and $\varphi_{\max}$. On the left eye panoramic image we can assign the angle φ_l to each pixel within the stripe: $\varphi_{\min} \leq \varphi_l \leq \varphi_{\max}$. After finding the corresponding point on the right eye panoramic image, we can evaluate the angle φ_r in the same manner, according to the position of the corresponding pixel within the stripe: $\varphi_{\min} \leq \varphi_r \leq \varphi_{\max}$. Now let us assume that we can still calculate the angle θ as in the basic system (see the next section to clear the issue of why we can assume this):

$$2\theta = dx \cdot \theta_0, \tag{2}$$

where dx is the absolute value of the difference between x coordinates of the corresponding points in the left eye panoramic image and in the right eye panoramic image, while θ_0 is the angle corresponding to one pixel column of the captured image and consequently the angle for which we have to move the robotic arm if we build the panoramic images from only one column of the captured image. Using analogy for this equation and having in mind that we are building the panoramic images from stripes, we can write the following equation (Figs. 1 and 4):

$$2\theta = \theta_l + \theta_r. \tag{3}$$

When we use one column instead of stripes then $\theta_l = \theta_r$ (Fig. 1), but this is not necessary true if we use stripes. In general these two values are different, but the property following from the equation

$$\frac{\theta_l}{\theta_r} = \frac{\varphi_l}{\varphi_r} \tag{4}$$

shows that the ratio of these two values is related to the angles φ (Fig. 4). The bigger φ_l gets, the bigger gets the corresponding θ_l. Now we can simply express θ_r and θ_l from Eqs. (2), (3) and (4) as:

$$\theta_r = \frac{dx \cdot \theta_0}{(1+\frac{\varphi_l}{\varphi_r})}$$
$$\theta_l = dx \cdot \theta_0 - \theta_r.$$

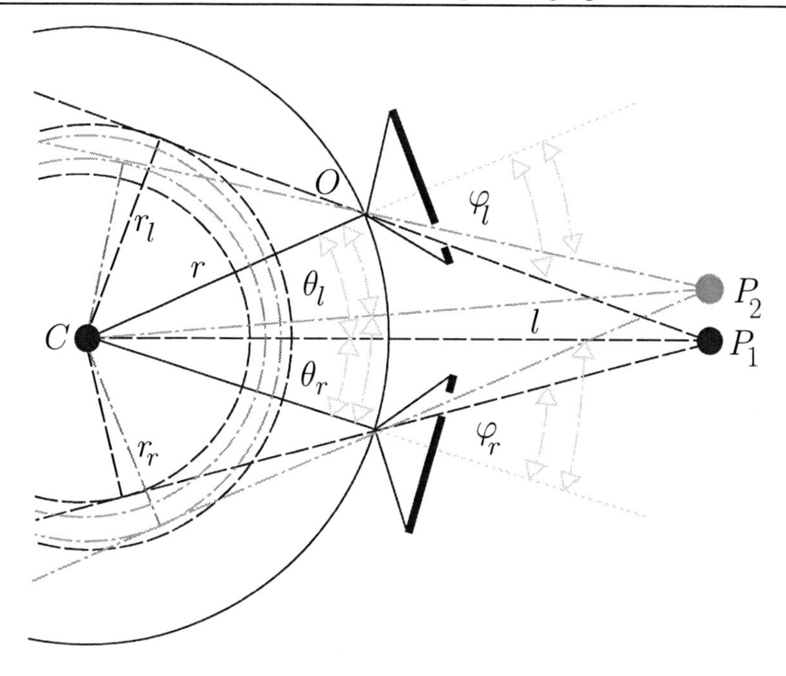

Figure 4. Angles θ_l and θ_r are related to angles φ_l and φ_r as presented in Eq. (4). Here the relationship is illustrated for two scene points.

We know that bigger φ brings bigger accuracy of the reconstruction process (see Sec. 3.). And since we would like to achieve the best accuracy possible, we take bigger φ from the two possible values (φ_l and φ_r) and associated θ and calculate the depth estimation using Eq. (1).

Note that the vertical reconstruction is addressed in the same way as in the basic system [13].

The same reconstruction principle could also be used when we do the reconstruction from the non-symmetric pair of panoramas build from only one pixel column of the captured images. The only difference in this case is that φ_l and φ_r take only the value given by the pixel column that contributes to each panoramic image. The experiment in Sec. 8.1. proves the correctness of the principle.

7. Epipolar Constraint

In the previous section we assumed that we can calculate the angle θ using Eq. (2). This equation holds if we do the reconstruction based on a symmetric pair of stereo panoramic images, which are made from one pixel column of the captured image. In this case we know that the epipolar lines are corresponding rows of the panoramic image (Sec. 3.).

The stripe is composed of columns, each of them with a different angle φ. This basically means that we are dealing in fact with non-symmetric cases, for which the epipolar lines are different from corresponding rows. But if we look at the situation from another viewpoint, we can establish the following: We are using symmetric stripes to build a stereo pair of panoramic images. If we lower the resolution of the captured image, we transform the stripe into a column. The symmetric stripes would become symmetric columns and we could again use the rows of the panoramic image as epipolar lines. The same conclusion can be drawn from the property of the viewing circle, which gets thicker if we use a stripe instead of a column.

We can see this fact already by observing the light rays forming the column or the stripe (Fig. 3) that contributes to the panoramic image. In both examples the viewing circle should be thicker as presented, because the column and the stripe are formed from the set of light rays, which intersect in the optical center of the camera.

Lower resolution also brings considerable decrease in the number of possible depth estimates [13], so the quality of results obtained from stripes should be better than the quality of results obtained from columns captured at a suitable lower resolution. For illustration: from around 140 possible estimates for *camera #1* in Sec. 8. to only around 10 estimates for the same camera at 14 times lower resolution – the width of the stripe that is transformed into a column W_s=14. It is also much harder to find corresponding points in such a low resolution image.

One very interesting property of the system is also that there is a possibility of reconstructing the scene from non-symmetric pairs of panoramas while still using the simple, horizontal epipolar constraint (see the experiment in Sec. 8.1.).

8. Experimental Results

In the experiments the following cameras were used (the width W of the captured images is 160 pixels in all experiments):

- *camera #1* with parameters: $\alpha = 34°$, $\beta = 25°$, $r = 30$ cm, $2\varphi_{\max} = 29.75°$, $2\varphi_{\min} = 2\varphi_{\max} - 2(W_s - 1)\frac{\alpha}{W}$ (note that $2\varphi = 29.9625°$ has been used when $W_s = 1$), $\theta_0 = W_s \cdot 0.205714°$,

- *camera #2* with parameters: $\alpha = 39.72°$, $\beta = 30.54°$, $r = 31$ cm, $2\varphi_{\max} = 34.755°$, $2\varphi_{\min} = 2\varphi_{\max} - 2(W_s - 1)\frac{\alpha}{W}$, $\theta_0 = W_s \cdot 0.257143°$,

- *camera #3* with parameters: $\alpha = 16.53°$, $\beta = 12.55°$, $r = 35.6$ cm, $2\varphi_{\max} = 14.46375°$, $2\varphi_{\min} = 2\varphi_{\max} - 2(W_s - 1)\frac{\alpha}{W}$, $\theta_0 = W_s \cdot 0.102857°$.

These values ensure similar number of possible depth estimates (Sec. 3.).

The normalized error of the estimated depth l in comparison to the actual distance d (in % of d) for the scene point i is given as:

$$ERR_{\%,i} = \frac{|l_i - d_i|}{d_i} \cdot 100.$$

Furthermore, the average error $AVG_{\%}$ (arithmetic mean) over n scene points is calculated. The second measure, which is in the results written right beside the first one ($AVG_{\%}$), is the standard deviation, which reveals how tightly all the various estimated depths are clustered around the average error in the set of data.

Correspondences for each feature point on the scene used in the evaluation have been determined with a *normalized correlation* procedure [6] and rechecked manually for consistency.

The real time sensor has been simulated by rotating one standard camera for the angle determined by the width of the stripe W_s. Since our final goal is to determine the usability of our system for mobile robot navigation, we have performed all the tests on real world images, so that the results reflect the applicability of implemented ideas in the real world.

8.1. Reconstruction from Non-symmetric Pairs of Panoramas

Experiment background: In Sec. 6. we have introduced the principle for reconstruction when we use symmetric stripes instead of symmetric columns. We have mentioned that the same principle could also be used when we do the reconstruction from the non-symmetric pairs of panoramas build from only one pixel column ($W_s = 1$) of the captured images.

And in Sec. 7. we have stated that we can still use the simple, horizontal epipolar constraint in some cases. These cases have been investigated by Shum and Szeliski [21] and they have concluded that even in the non-symmetric cases the epipolar constraint is sufficiently close to the horizontal epipolar constraint if either r/l or φ are small. They have even presented the approximate numbers: $r/l \leq 0.\overline{6}$, $2\varphi \leq 30°$. In the example given here, we satisfied both criteria.

In our case in this section the corresponding points lie in the same image row determined by the epipolar constraint. The results were obtained with *camera #1*.

Results: The comparison of results obtained by processing symmetric pair of panoramas and non-symmetric pair of panoramas is presented in Tab. 1.

Conclusion: We can see that the results are similar. Thus, the experiment confirms that the suggested reconstruction principle is correct.

8.2. Reconstruction from Stripe Panoramas

Experiment background: We want to determine the influence of the width of the stripe (W_s) that contributes to the panoramic image on the reconstruction accuracy ($\text{AVG}_{\%}$).

The results were obtained with *camera #1*.

Results: The results for four different widths of the stripes (W_s) are given in Tab. 2. Note that the width of the captured images from which the panoramic images have been constructed is 160 pixels.

Table 1. The results obtained by processing symmetric pair of panoramas ($2\varphi = 29.9625°$) and non-symmetric pair of panoramas ($2\varphi_l = 29.9625°$, $2\varphi_r = 3.6125°$).

feature	d [cm]	symmetric pair $l(\alpha, \beta)$ [cm]	symmetric pair $l(\alpha, \beta) - d$ [cm (% of d)]	non-symmetric pair $l(\alpha, \beta)$ [cm]	non-symmetric pair $l(\alpha, \beta) - d$ [cm (% of d)]
1	165.0	162.3	-2.7 (-1.6%)	155.3	-9.7 (-5.9%)
2	119.0	118.0	-1.0 (-0.9%)	112.1	-6.9 (-5.8%)
3	128.0	133.7	5.7 (4.4%)	131.4	3.4 (2.7%)
4	126.5	125.6	-0.9 (-0.7%)	124.4	-2.1 (-1.7%)
5	143.0	146.7	3.7 (2.6%)	146.8	3.8 (2.6%)
6	143.0	151.9	8.9 (6.2%)	146.8	3.8 (2.7%)
7	142.5	152.7	10.2 (7.2%)	147.6	5.1 (3.6%)
8	136.5	141.0	4.5 (3.3%)	137.6	1.1 (0.8%)
9	104.5	106.8	2.3 (2.2%)	103.0	-1.5 (-1.4%)
10	81.7	79.6	-2.1 (-2.5%)	79.7	-2.0 (-2.5%)
11	84.5	80.6	-3.9 (-4.6%)	81.9	-2.6 (-3.1%)
12	83.5	82.7	-0.8 (-0.9%)	82.4	-1.1 (-1.3%)
13	97.0	94.9	-2.1 (-2.2%)	91.2	-5.8 (-6.0%)
14	110.0	114.9	4.9 (4.5%)	107.3	-2.7 (-2.5%)
15	180.0	191.1	11.1 (6.2%)	192.4	12.4 (6.9%)
16	124.5	129.9	5.4 (4.3%)	124.9	0.4 (0.3%)
17	132.5	132.4	-0.1 (-0.1%)	130.2	-2.3 (-1.8%)
18	134.5	136.6	2.1 (1.5%)	130.2	-4.3 (-3.2%)
19	113.0	109.4	-3.6 (-3.2%)	107.5	-5.5 (-4.8%)
20	125.0	121.6	-3.4 (-2.8%)	117.8	-7.2 (-5.7%)
21	130.0	128.8	-1.2 (-1.0%)	123.8	-6.2 (-4.8%)
		$AVG_{\%}$=3% ± 2%		$AVG_{\%}$=3.3% ± 1.9%	

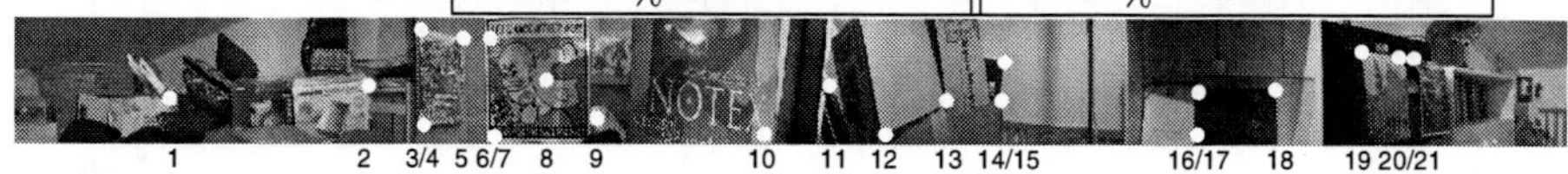

Table 2. The results obtained with four different widths of the stripes (W_s).

feature	d [cm]	W_s=2 $l(\alpha,\beta)$ [cm]	W_s=2 $l(\alpha,\beta)-d$ [cm (% of d)]	W_s=6 $l(\alpha,\beta)$ [cm]	W_s=6 $l(\alpha,\beta)-d$ [cm (% of d)]	W_s=14 $l(\alpha,\beta)$ [cm]	W_s=14 $l(\alpha,\beta)-d$ [cm (% of d)]	W_s=29 $l(\alpha,\beta)$ [cm]	W_s=29 $l(\alpha,\beta)-d$ [cm (% of d)]
1	165.0	165.8	0.8 (0.5%)	162.9	-2.1 (-1.3%)	130.1	-34.9 (-21.1%)	96.3	-68.7 (-41.6%)
2	119.0	114.1	-4.9 (-4.1%)	103.2	-15.8 (-13.3%)	117.4	-1.6 (-1.3%)	113.0	-6.0 (-5.0%)
3	128.0	136.3	8.3 (6.5%)	141.1	13.1 (10.2%)	133.0	5.0 (3.9%)	163.4	35.4 (27.7%)
4	126.5	127.7	1.2 (0.9%)	119.0	-7.5 (-5.9%)	132.5	6.0 (4.8%)	128.9	2.4 (1.9%)
5	143.0	145.0	2.0 (1.4%)	143.3	0.3 (0.2%)	117.6	-25.4 (-17.8%)	225.2	82.2 (57.5%)
6	143.0	154.7	11.7 (8.2%)	143.3	0.3 (0.2%)	117.5	-25.5 (-17.8%)	89.7	-53.3 (-37.3%)
7	142.5	155.5	13.0 (9.1%)	161.7	19.2 (13.5%)	179.3	36.8 (25.8%)	179.8	37.3 (26.2%)
8	136.5	144.0	7.5 (5.5%)	126.9	-9.6 (-7.0%)	111.3	-25.2 (-18.5%)	177.4	40.9 (29.9%)
9	104.5	103.5	-1.0 (-0.9%)	101.6	-2.9 (-2.8%)	112.8	8.3 (7.9%)	97.1	-7.4 (-7.1%)
10	81.7	80.1	-1.6 (-2.0%)	76.7	-5.0 (-6.1%)	75.4	-6.3 (-7.7%)	57.3	-24.4 (-29.9%)
11	84.5	79.6	-4.9 (-5.8%)	81.1	-3.4 (-4.0%)	74.9	-9.6 (-11.3%)	55.0	-29.5 (-34.9%)
12	83.5	83.3	-0.2 (-0.3%)	81.6	-1.9 (-2.2%)	83.2	-0.3 (-0.3%)	58.8	-24.7 (-29.6%)
13	97.0	97.4	0.4 (0.4%)	95.5	-1.5 (-1.6%)	78.6	-18.4 (-19.0%)	74.1	-22.9 (-23.6%)
14	110.0	113.6	3.6 (3.2%)	107.8	-2.2 (-2.0%)	117.6	7.6 (6.9%)	101.7	-8.3 (-7.6%)
15	180.0	182.0	2.0 (1.1%)	165.9	-14.1 (-7.8%)	203.2	23.2 (12.9%)	96.4	-83.6 (-46.4%)
16	124.5	128.2	3.7 (3.0%)	125.8	1.3 (1.0%)	142.0	17.5 (14.1%)	138.6	14.1 (11.3%)
17	132.5	127.7	-4.8 (-3.6%)	131.6	-0.9 (-0.7%)	150.8	18.3 (13.8%)	148.2	15.7 (11.9%)
18	134.5	134.9	0.4 (0.3%)	131.6	-2.9 (-2.1%)	97.8	-36.7 (-27.3%)	119.7	-14.8 (-11.0%)
19	113.0	113.1	0.1 (0.1%)	111.0	-2.0 (-1.8%)	101.9	-11.1 (-9.8%)	97.1	-15.9 (-14.1%)
20	125.0	126.5	1.5 (1.2%)	118.6	-6.4 (-5.1%)	124.5	-0.5 (-0.4%)	120.0	-5.0 (-4.0%)
21	130.0	127.2	-2.8 (-2.2%)	137.4	7.4 (5.7%)	140.9	10.9 (8.4%)	137.8	7.8 (6.0%)
		$AVG_{\%}$=2.9% ± 2.7%		$AVG_{\%}$=4.5% ± 4%		$AVG_{\%}$=11.9% ± 7.9%		$AVG_{\%}$=22.1% ± 15.8%	

1 2 3/4 5 6/7 8 9 10 11 12 13 14/15 16/17 18 19 20/21

Conclusion: As expected, the accuracy deteriorates with wider stripes. Though, if we compare the results obtained with W_s=2 and W_s=6 with the results obtained with W_s=1 (e.g. Tab. 1), we see that the results are still very good if not very similar.

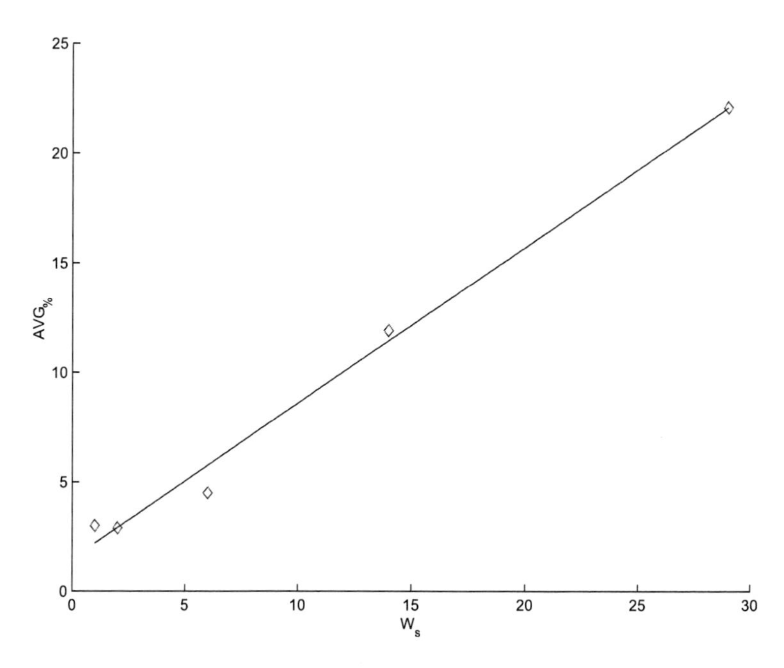

Figure 5. The relation between $\text{AVG}_{\%}$ and W_s ($\text{AVG}_{\%}(W_s)$) can be well approximated with a linear function. The diamonds represent the data involved in the approximation.

The graph in Fig. 5 shows that the relation between $\text{AVG}_{\%}$ and W_s ($\text{AVG}_{\%}(W_s)$) can be well approximated with a linear function, although we have fitted a second-order polynomial on the (sparse) data.

Another fact is that with the wider stripe more scene points are not captured in the panoramic images (Sec. 4.1.), which means that the number of pixels in the left eye panoramic image that do not have the corresponding point in the right eye panoramic image increases. So, the generated panoramic images differ more and more from the ideal case (W_s=1) as we are increasing the width of the stripe W_s. In the ideal case each scene point in the left eye panoramic image has its corresponding point in the right eye panoramic image; exceptions are of course related to occlusions. – We performed an experiment in which we have used the *camera #1* (Sec. 8., W_s=1) to capture a banner written all over with text of different sizes, styles and fonts positioned at approximately 2 cm from the camera lens (on each location of the camera on the camera path). We were unable to prove from the human perceptual point of view the hypothesis that some scene points in this case are not captured. Really small deviations of the generated panoramic images from the real banner could as well be attributed to the fact that the images are discrete (for example: we cannot take a half of a pixel). On the other hand, the hypothesis is confirmed if stripes ($W_s > 1$) are used (Fig. 2).

8.3. Reconstruction from Stripe Panoramas — Different Room

Experiment background: We want to see if we can achieve similar results as in Sec. 8.2., using the same camera (*camera #1*) in a different room?
Results: The results obtained for two most interesting stripe widths (according to Tab. 2) in the different room are presented in Tab. 3.
Conclusion: The overall results are a bit different, especially the results for W_s=14, though still similar. But this was also expected since the results for W_s=1 ($AVG_{\%}$=4.5% $\pm$ 4.5%) are also a bit worse compared to the results obtained in a previous room (the symmetric pair in Tab. 1). Small differences in results are expected, since each room has its own shape, i.e. the depth distribution around the center of the system is different. And we know how this influences the accuracy, while we are limited with the number of possible depth estimates, which are approximations of the real distances (Sec. 3.).

8.4. Reconstruction from Stripe Panoramas — Different Cameras

Experiment background: We want to see how is the reconstruction error related to different cameras used in the same room, when the width of the stripe is constant for all cameras? As written in Sec. 8. the results were obtained at similar number of possible depth estimates.
Results: The comparison of results for the width of the stripe W_s=14 for three different cameras is given in Tab. 4. Note that for features marked 3, 5, 6, 7, 15, 19, 20 and 21 the real distance d in case of *camera #3* is different from the presented one. The reason for this lies in the vertical view angle of the camera β, which is smaller in comparison to other two cameras. This means that some marked feature points are not seen in the panoramic images generated with the

Table 3. The results obtained in the different room, but with the same camera as in Sec. 8.2..

feature	d [cm]	W_s=6 $l(\alpha,\beta)$ [cm]	W_s=6 $l(\alpha,\beta)-d$ [cm (% of d)]	W_s=14 $l(\alpha,\beta)$ [cm]	W_s=14 $l(\alpha,\beta)-d$ [cm (% of d)]
1	63.2	59.0	-4.2 (-6.6%)	57.0	-6.2 (-9.8%)
2	51.5	48.8	-2.7 (-5.3%)	54.3	2.8 (5.5%)
3	141.0	161.3	20.3 (14.4%)	196.7	55.7 (39.5%)
4	142.0	143.6	1.6 (1.1%)	246.6	104.6 (73.7%)
5	216.0	213.3	-2.7 (-1.2%)	225.7	9.7 (4.5%)
6	180.0	209.2	29.2 (16.2%)	184.9	4.9 (2.7%)
7	212.0	200.4	-11.6 (-5.5%)	167.0	-45.0 (-21.2%)
8	49.0	44.8	-4.2 (-8.6%)	48.6	-0.4 (-0.9%)
9	49.0	45.5	-3.5 (-7.2%)	45.5	-3.5 (-7.2%)
10	97.0	95.7	-1.3 (-1.3%)	90.6	-6.4 (-6.6%)
11	129.5	132.6	3.1 (2.4%)	102.1	-27.4 (-21.1%)
12	134.0	139.8	5.8 (4.3%)	162.8	28.8 (21.5%)
13	119.0	122.7	3.7 (3.1%)	117.9	-1.1 (-0.9%)
14	156.0	163.0	7.0 (4.5%)	138.2	-17.8 (-11.4%)
15	91.0	94.0	3.0 (3.3%)	81.7	-9.3 (-10.2%)
16	97.7	99.6	1.9 (1.9%)	96.9	-0.8 (-0.8%)
17	111.0	110.9	-0.1 (-0.1%)	169.4	58.4 (52.6%)
18	171.5	191.1	19.6 (11.4%)	147.3	-24.2 (-14.1%)
19	171.5	165.8	-5.7 (-3.3%)	170.4	-1.1 (-0.6%)
		AVG$_{\%}$=5.4% ± 4.5%		AVG$_{\%}$=16% ± 19.6%	

camera #3, so we have chosen a nearby features with similar distances. By all means, in the calculations we have used the correct distances.

Conclusion: The results show that with bigger horizontal view angle α and at constant width of the stripe W_s (> 1) the reconstruction accuracy deteriorates:

Table 4. The results obtained with three different cameras, while the width of the stripe W_s=14.

feature	d [cm]	*camera #1* $l(\alpha,\beta)$ [cm]	*camera #1* $l(\alpha,\beta)-d$ [cm (% of d)]	*camera #2* $l(\alpha,\beta)$ [cm]	*camera #2* $l(\alpha,\beta)-d$ [cm (% of d)]	*camera #3* $l(\alpha,\beta)$ [cm]	*camera #3* $l(\alpha,\beta)-d$ [cm (% of d)]
1	165.0	130.1	-34.9 (-21.1%)	178.7	13.7 (8.3%)	144.6	-20.4 (-12.3%)
2	119.0	117.4	-1.6 (-1.3%)	167.0	48.0 (40.4%)	130.0	11.0 (9.3%)
3	128.0	133.0	5.0 (3.9%)	120.9	-7.1 (-5.5%)	118.1	-7.4 (-5.9%)
4	126.5	132.5	6.0 (4.8%)	99.0	-27.5 (-21.7%)	95.5	-31.0 (-24.5%)
5	143.0	117.6	-25.4 (-17.8%)	127.4	-15.6 (-10.9%)	124.6	-16.4 (-11.6%)
6	143.0	117.5	-25.5 (-17.8%)	231.5	88.5 (61.9%)	114.3	-27.2 (-19.2%)
7	142.5	179.3	36.8 (25.8%)	134.8	-7.7 (-5.4%)	154.5	12.5 (8.8%)
8	136.5	111.3	-25.2 (-18.5%)	183.0	46.5 (34.1%)	119.2	-17.3 (-12.7%)
9	104.5	112.8	8.3 (7.9%)	106.7	2.2 (2.1%)	96.8	-7.7 (-7.4%)
10	81.7	75.4	-6.3 (-7.7%)	95.9	14.2 (17.4%)	78.8	-2.9 (-3.5%)
11	84.5	74.9	-9.6 (-11.3%)	69.5	-15.0 (-17.7%)	88.8	4.3 (5.0%)
12	83.5	83.2	-0.3 (-0.3%)	69.7	-13.8 (-16.5%)	75.1	-8.4 (-10.0%)
13	97.0	78.6	-18.4 (-19.0%)	93.3	-3.7 (-3.8%)	102.3	5.3 (5.5%)
14	110.0	117.6	7.6 (6.9%)	98.8	-11.2 (-10.2%)	108.2	-1.8 (-1.6%)
15	180.0	203.2	23.2 (12.9%)	197.1	17.1 (9.5%)	174.8	3.8 (2.2%)
16	124.5	142.0	17.5 (14.1%)	108.3	-16.2 (-13.0%)	118.1	-6.4 (-5.2%)
17	132.5	150.8	18.3 (13.8%)	167.0	34.5 (26.0%)	123.6	-8.9 (-6.7%)
18	134.5	97.8	-36.7 (-27.3%)	153.9	19.4 (14.5%)	102.0	-32.5 (-24.2%)
19	113.0	101.9	-11.1 (-9.8%)	126.3	13.3 (11.8%)	100.4	-7.6 (-7.0%)
20	125.0	124.5	-0.5 (-0.4%)	168.3	43.3 (34.6%)	129.9	10.9 (9.2%)
21	130.0	140.9	10.9 (8.4%)	183.8	53.8 (41.4%)	102.1	-22.9 (-18.3%)
		$AVG_{\%}$=11.9% ± 7.9%		$AVG_{\%}$=19.4% ± 15.3%		$AVG_{\%}$=10% ± 6.6%	

1 2 3/4 5 6/7 8 9 10 11 12 13 14/15 16/17 18 19 20/21

$$\alpha_{camera\#3} < \alpha_{camera\#1} < \alpha_{camera\#2}$$
$$\Longleftrightarrow$$
$$\mathrm{AVG}_{\%_{camera\#3}} < \mathrm{AVG}_{\%_{camera\#1}} < \mathrm{AVG}_{\%_{camera\#2}}.$$

The result in Tab. 3 for W_s=14, gained with *camera #1*, but in the different room, also satisfies this relation.

Bigger horizontal view angle α means that the lens distortion in captured images is more obvious. Thus, this fact in relation with the above conclusion suggests that by using undistorted images, we could obtain better reconstruction accuracy. But the overall reconstruction accuracy obtained using undistorted sequence is very similar to that obtained using distorted sequence. This is due to the fact that we can see that the problem with missing scene points in the panoramic images (Sec. 4.1.) remains.

So, one of the reasons for the lower reconstruction accuracy is again the fact that there are parts of the scene which are not captured in the panoramic image. This means that the number of pixels in the left eye panoramic image that do not have the corresponding point in the right eye panoramic image increases with bigger horizontal view angle α at constant width of the stripe W_s, while the shift angle θ_0 between two successively positioned cameras is bigger for bigger α (Sec. 8.).

In Sec. 3. we have concluded that similar overall accuracy can be achieved if we use different cameras, while the width of the stripe W_s=1. If we relate this conclusion to the main conclusion of this section (with bigger horizontal view angle α and at constant width of the stripe W_s (> 1) the reconstruction accuracy deteriorates), we can conclude that the approximation presented in Fig. 5 gets steeper with increasing horizontal view angle α.

9. Conclusions

This chapter evaluated our idea for the real time extension of the basic mosaic-based panoramic depth imaging system.

The presented theory and results suggest that the new sensor could be used for real time capturing of panoramic depth images and consequently for autonomous navigation of a mobile robot in a room. But this statement is unfortunately highly related to the application of the system and its demanded reconstruction accuracy. If in Sec. 3. we could conclude that by all means the basic system can be used for robot localization and navigation in a room, we are limited with the number of constraints when it comes to the real time sensor. On the other hand, assumptions made in this chapter have proved to be correct and revealed some other interesting properties of the system.

The main conclusions made in this chapter are:

- Building panoramic images from wider stripes than only one-pixel column brings faster execution of the building process and worse quality of the generated panoramas (Sec. 4.).

- The stripes that contribute to the panoramic image do not cover all the scene (Sec. 4.1.) The wider is the stripe, more scene points are not captured in the panoramic image (the number of points without correspondences increases (Sec. 8.2.)) By using smaller r the gaps are smaller, which means that with smaller r we cover more scene points. The same goes for different values of α. This holds also for one-pixel column stripes.

- If we want to capture the parallax effect, which is essential for the reconstruction, we have to accept the fact that not all the scene points are captured (Sec. 4.1.)

- If we could have a camera on each position on the circular path, we could build the panoramic image in real time and achieve the same quality of results as in case of the basic system, but in practice we should make a compromise between the width of the stripes, the number of cameras in respect to their parameters (e.g. α and external dimensions), the value of radius r and the needed accuracy of the system (Sec. 5.)

- The stereo reconstruction procedures along with the epipolar constraint have been proved correct for both investigated cases: the basic sensor and the real time sensor (Secs. 6., 7. and 8.)

- Using images with resolution 160×120 represents a good compromise between overall time complexity of the system and its accuracy (Secs. 3. and 5.)

- Even in some cases of non-symmetric pairs of panoramas we can use our reconstruction procedure (Sec. 8.1.)

- The reconstruction accuracy deteriorates approximately linearly with wider stripes (Sec. 8.2.)

- We can achieve similar reconstruction accuracy with panoramas build from stripes ($W_s > 1$) with fixed size W_s of the captured images in different rooms at fixed horizontal view angle α (Sec. 8.3.), but with bigger horizontal view angle α and at constant width of the stripe W_s (> 1) the reconstruction accuracy deteriorates (Sec. 8.4.)

- In general, by undistorting the image sequence and building panoramas from stripes ($W_s > 1$), we do not obtain better reconstruction accuracy (Sec. 8.4.)
- The approximate linear dependency of accuracy on the width of the stripes gets steeper with increasing horizontal view angle α (Sec. 8.4.)

All this is true for the cameras used in the chapter, while for really wide angle cameras some conclusions perhaps demand further investigation in direction presented by the conclusion.

Acknowledgment

This work was supported by the Ministry of Education, Science and Sport of Republic of Slovenia (programme Computer Vision).

References

[1] Bakstein H., Pajdla T.: Panoramic Mosaicing with a 180° Field of View Lens. Proc. *IEEE Workshop on Omnidirectional Vision*, Copenhagen, Denmark, June, 2002, 60–67.

[2] Bakstein H., Pajdla T.: Ray space volume of omnidirectional 180°×360° images. Proc. *Computer Vision Winter Workshop (CVWW)*, Valtice, Czech Republic, February 3–6, 2003, 39–44.

[3] Benosman R., Kang S. B. (Eds.): *Panoramic Vision: Sensors, Theory and Applications*. Springer-Verlag, New York, USA, 2001.

[4] Bouguet J.-Y.: Camera Calibration Toolbox for Matlab. California Institute of Technology. Available at: http://www.vision.caltech.edu/bouguetj/calib_doc/index.html .

[5] Chen S.: Quicktime VR — an image-based approach to virtual environment navigation. Proc. *ACM SIGGRAPH*, Los Angeles, USA, August 6–11, 1995, 29–38.

[6] Faugeras O.: *Three-Dimensional Computer Vision: A Geometric Viewpoint*. MIT Press, Cambridge, Massachusetts, London, England, 1993.

[7] Feldman D., Zomet A., Weinshall D., Peleg S.: New view synthesis with non-stationary mosaicing. Proc. *Computer Vision / Computer Graphics Collaboration for Model-based Imaging, Rendering, image Analysis and Graphical special Effects (MIRAGE)*, INRIA Rocquencourt, France, March 10–11, 2003, 48–56.

[8] Gupta R., Hartley R. I.: Linear pushbroom cameras. *IEEE Trans. PAMI*, **19**(9), 963–975, 1997.

[9] Huang F., Pajdla T.: Epipolar geometry in concentric panoramas. *Technical Report CTU-CMP-2000-07*, Center for Machine Perception, Czech Technical University, Prague, Czech Republic, 2000. Available at: ftp://cmp.felk.cvut.cz/pub/cmp/articles/pajdla/Huang-TR-2000-07.ps.gz .

[10] Huang F., Wei S. K., Klette R.: Geometrical Fundamentals of Polycentric Panoramas. Proc. *IEEE ICCV*, Vancouver, Canada, July 9–12, 2001, I:560–565.

[11] Ishiguro H., Yamamoto M., Tsuji S.: Omni-directional stereo. *IEEE Trans. PAMI*, **14**(2), 257–262, 1992.

[12] Peer P., Solina F.: Panoramic Depth Imaging: Single Standard Camera Approach. *International Journal of Computer Vision*, **47**(1/2/3), 149–160, 2002.

[13] Peer P., Solina F.: Multiperspective panoramic depth imaging, chapter in *Computer Vision and Robotics*, John X. Liu, Ed. Nova Science Publishers, New York, pp. 135–188, 2006.

[14] Peer P., Solina F.: Where physically is the optical center? *Pattern recognition letters*, **27**(10), pp. 1117–1121, 2006.

[15] Peleg S., Ben-Ezra M.: Stereo panorama with a single camera. Proc. *IEEE CVPR*, Fort Collins, USA, June 23–25, 1999, I:395–401.

[16] Peleg S., Pritch Y., Ben-Ezra M.: Cameras for stereo panoramic imaging. Proc. *IEEE CVPR*, Hilton Head Island, USA, June 13–15, 2000, I:208–214.

[17] Peleg S., Rousso B., Rav-Acha A., Zomet A.: Mosaicing on adaptive manifolds. *IEEE Trans. PAMI*, **22**(10), 1144–1154, 2000.

[18] Peleg S., Ben-Ezra M., Pritch Y.: Omnistereo: Panoramic Stereo Imaging. *IEEE Trans. PAMI*, **23**(3), 279–290, 2001.

[19] Rademacher P., Bishop G.: Multiple-center-of-projection images. *Computer Graphics (ACM SIGGRAPH)*, Orlando, USA, July 19–24, 1998, 199–206.

[20] Shimada D., Tanahashi H., Kato K., Yamamoto K.: Extract and Display Moving Object in All Direction by Using Stereo Omnidirectional System (SOS). Proc. *IEEE International Conference on 3-D Digital Imaging and Modeling*, Quebec City, Canada, May 28 – June 1, 2001, 42–47.

[21] Shum H. Y., Szeliski R.: Stereo Reconstruction from Multiperspective Panoramas. Proc. *IEEE ICCV*, Kerkyra, Greece, September 20–25, 1999, I:14–21.

[22] Shum H. Y., Kalai A., Seitz S. M.: Omnivergent Stereo. Proc. *IEEE ICCV*, Kerkyra, Greece, September 20–25, 1999, I:22–29.

[23] Sivic J.: Geometry of Concentric Multiperspective Panoramas. *M.Sc. Thesis*, Center for Machine Perception, Czech Technical University, Prague, Czech Republic, 2002.

[24] Sun C., Peleg S.: Fast Panoramic Stereo Matching Using Cylindrical Maximum Surfaces. *IEEE Trans. SMC – part B: Cybernetics*, **34**(1), 760–765, 2004.

[25] Svoboda T., Pajdla T.: Epipolar Geometry for Central Catadioptric Cameras. *International Journal of Computer Vision*, **49**(1), 23–37, 2002.

[26] Tanahashi H., Yamamoto K., Wang C., Niwa Y.: Development of a Stereo Omnidirectional Imaging System (SOS). Proc. *IEEE International Conference on Industrial Electronics, Control and Instrumentation*, Nagoya, Japan, October 22–28, 2000, 289–294.

[27] Tanahashi H., Shimada D., Yamamoto K., Niwa Y.: Acquisition of Three-Dimensional Information in Real Environment By Using Stereo Omnidirectional System (SOS). Proc. *IEEE International Conference on 3-D Digital Imaging and Modeling*, Quebec City, Canada, May 28 – June 1, 2001, 365–371.

[28] Wood D., Finkelstein A., Hughes J., Thayer C., Salesin D.: Multiperspective panoramas for cel animation. *Computer Graphics (ACM SIGGRAPH)*, Los Angeles, USA, August 3–8, 1997, 243–250.

[29] Zomet A., Feldman D., Peleg S., Weinshall D.: Mosaicing New Views: The Crossed-Slits Projection. *IEEE Trans. PAMI*, **25**(6), 741–754, 2003.

In: New Robotics Research ISBN 978-1-60741-093-5
Editors: E.D. Wagner et al, pp. 107-130

Chapter 6

REGISTRATION TECHNIQUES FOR IMAGE GUIDED ROBOTIC SURGERY

Aleksandra Popovic* and Karen I. Trovato
Philips Research North America

Abstract

The ultimate objective of surgical robotic systems is to carry out a preoperative plan accurately with precision. A surgical robot is part of a larger system, comprised of imaging device(s), position tracking and various instruments. For effective and accurate cooperation between these systems, they must be registered to the patient.

The goal of intraoperative registration is to establish a relationship between the frame of reference of the robotic system and the preoperative plan, typically generated in the coordinate system of the imaging device. This step has crucial impact on the overall accuracy, since registration inaccuracies are significantly higher than those of the robots mechanical and control systems. The main causes of registration inaccuracy are limitations of the tracking system, including environmental interference, difficulty accessing anatomical landmarks, such as in minimally invasive surgery, and inherent inaccuracies of the registration algorithm, including improper selection of registration fiducials.

The first part of this chapter provides an analysis of various tracking techniques (optical, electromagnetic, ultrasonic, and laser-based) in different environments and surgical fields. The second part of this chapter

*E-mail address: aleksandra.popovic@philips.com

gives an overview of registration algorithms, with an analysis of theoretical registration error-distribution followed by practical design considerations for an accurate registration set-up. The final part of this chapter summarizes the registration methods used for commercial and research surgical robots.

Key Words: Surgical robotics, Navigation, Registration, Tracking

1. Image Guided Surgery and Surgical Robots

As robots become more accepted in operating rooms, the issue of information integration becomes the central problem in minimally invasive surgery. Intraoperative techniques have made it possible to assess immediate impact during the surgery, allowing flexibility and adaptability of the therapy. These developments may lead to more personalized healthcare with therapy individually customized for each patient. The integration and utilization of patient data is a challenge, given the amount of information available. In the field of medical image processing, a great effort has been made to propose methods to analyze image information and register images taken from different modalities or at different times. Surgical planning systems utilize this information to provide patient specific therapy planning. The challenge is to transfer the plan to the operating theater. Image guidance systems provide navigational help to the surgeon, by establishing a relationship between patient, surgical instruement(s), preoperative/intraoperative images the plan and other data, such as anatomical atlases. Surgical robots are emerging as assistance systems to allow higher accuracy and precision, improve efficacy and safety. Although development of surgical robots did not always follow development of imaging techniques, integration of robotics in the work-flow of image guided surgery has been a long lasting goal. Surgical robotic systems can be coarsely classified by increasing levels of autonomy:

- **Teleoperated** robots perform remote manipulation, in a master-slave configuration; a surgeon has full control and must direct each robotic movement. The main benefit of teleoperated robots is scaled motion and sight, enabling precise surgery in small areas. This improves dexterity of the end-effector and can reduce the invasiveness of the procedure. Moreover, teleoperated robots allow manipulation in small and hazardous environments, such as inside an MRI or CT scanner. Finally, the master-slave configuration might be used for the remote/internet surgery.

- **Interactive** robots provide assistance to surgeon, by imposing active (programmable) or passive (e.g. kinematic) constraints to surgical gestures. Robots with programmable constraints, such as a programmed robotic path, where surgeon controls force, velocity or depth, are sometimes referred to as synergistic robots. As compared to teleoperated systems, interactive systems offer a direct translation of surgeons movements as well as a direct feedback, reducing latency in the systems. Interactive system have improved accuracy, assuming that mechanical precision of the robot is better than the hand-eye coordination of a surgeon.

- **Active** robots perform a portion of the surgery, based on the plan that is approved by the surgeon. The systems often rely on precise and accurate motion and high-speed and precise tracking to achieve repeatable clinical results.

In the context of image guidance, active and interactive robotic systems are of particular interest because they usually utilize quantitative preoperative planning more often than to teleoperated robots. Image guidance is already used for telerobotic systems, however images are provided directly to the surgeon who mentally performs the matching between the preoperative images and the intraoperative images of the patient. This chapter focuses on the technologies required to create automatic or semi-automatic control for robotic systems. One of the key challenges is how to keep track of the target and tool locations during a procedure. The process to establish a relationship between the patient, the robot, and the preoperative images/plan is referred to as *registration*, or, sometimes in the case of registration of robot to patient, as *calibration*. In this chapter, the term **registration** will be used, to avoid a confusion with an initialization of the robot.

In the following sections, the background in tracking techniques and registration algorithms will be given, with an overview of the registration approaches used in the current commercial and research surgical robotic systems.

2. Tracking Techniques

Spatial position and orientation measurement is one of the enabling techniques for image guided robotics surgery because they provide live (near real-time) data. Tracking systems can be classified by the type of sensors used:

- Mechanical
- Optical, using either:
  - Infrared (IR) light to track:
    - passive retro-reflective elements, or
    - active light emitting diodes (LED)
  - Video processing
- (Electro)magnetic
- Ultrasonic, using either:
  - time-of-flight (TOF), or
  - phase coherence
- MR tracking using microcoils

Table 2. gives an overview of some tracking systems with their main characteristics.

Mechanical tracking systems measure tip location and orientation through a direct mechanical coupling of the device and a measurement point. This is typically, a 6 Degrees-of-Freedom (DOF) arm equipped with encoders on all joints, measuring a relative position and orientation change with respect to the robot base (for example MicroScribe MX, Immersion, CA, USA or Phantom R Omni, SensAble, MA, USA). Alternatively, shape measurement sensors, such as ShapeTape™(Measurand Inc., Canada), use optical fiber bend sensors to detect shape deformation throughout the entire shape.

Optical IR-based tracking systems are the leading position and orientation measurement component in the image guided and robotic surgical systems. They use triangulation from multiple cameras to localize IR-LED or IR reflective markers in the 6D space (3D position and 3 axis orientation). The markers are set in a known rigid configuration, so that they can be identified uniquely.

Each marker configuration is referred to as a *rigid body*. In case of passive markers (retro-reflective spheres), a rigid body must have a unique configuration to allow distinction, after characterized. Despite wide belief that active markers are more accurate and reliable than passive markers, there are studies indicating that they have comparable performance [37]. An alternative to IR-based optical systems (such as Polaris Spectra, NDI, Canada), videometric systems (Microtracker, Claron Technology Inc., Canada), use stereoscopic vision to detect high contrast visual patterns attached to tracked objects, in real time.

Electromagnetic tracking systems use an AC (Aurora, Northern Digital, Canada) or a pulse DC (Flock of Birds, Ascension, VT) magnetic field to measure position and orientation of the sensors, by measuring induced current in the sensor coils (for example, three orthogonal coils for a 6DOF measurement). The field strength in each of the orthogonal components can be computed from the induced currents. As the initial field strength is known as well as the nature of the field strength decrease (depending on distance and angular displacement), it is possible to compute the 3D position of the coils. Orientation can be computed from the phase difference of the induced currents in the three coils.

Standard ultrasonic tracking systems use acoustic waves to estimate distance. Ultrasound trackers use time-of-flight (TOF) or phase coherence techniques to determine position and orientation [32]. In image guidance tracking, ultrasonic systems are often combined with optical tracking, for example with A-mode ultrasound [26], or using image matching [24].

MR tracking using microcoils is a novel technique to track objects inside of the MR scanner's magnetic field. The microcoils can be passive, independent resonant circuits, detected by image processing (e.g. [27]), or active, connected to receive channel of the scanner, detecting position by applying magnetic gradient (e.g. [35]).

Selection of a tracking system depends on the application, specifically on the intraoperative set-up and the accuracy/precision requirements. Mechanical systems demonstrate very good accuracy, precision, and stability, however a mechanical arm can track only one point at a time, which is a limiting factor for many applications. Furthermore, mechanical arms require direct contact with the tracked point, requiring access space. An advantage of mechanical position sensing in image guided robotic surgery is that a robot can act as a mechanical tracking device, that might be used to directly and accurately register itself to the patient.

Manufacturer		Type	Accuracy[1] [mm]	Field of measurement [mm]	Number of tools	Physical size	Update rate Hz
	Mechanical						
Immersion	Microscribe MX	Arm	0.07	1270	1		
SensAble	Phantom R Omni	Haptic	0.05	160 x 120 x 70	1	168 x 2003 (footprint)	
Measurerand	ShapeTape	Optical sensors	0.3;0.5°	tape length	1	1.3 x 13 x 1800 (tape)	110
	Optical						
NDI	Certus	Active tools	0.15	5.5 x 2.6 x 3.6	170	1126 x 200 x 161	4600
NDI	Polaris Spectra	Hybrid	0.25/0.3	1.45 x 1.3 x 1.6[2]	15	613 x 104 x 86	60
NDI	Polaris Vicra	Hybrid	0.25/0.3		6	273 x 69 x 69	20
Claron	MicroTracker 2 Sx60	Videometric	0.25	1.15 x 0.70 x 0.55	unlimited	157 x 36 x 47 mm	48
Acension	LaserBIRD 2	Laser	0.5°	0.15 to 1.83 m	1	32 cm x 9 cm x 4 cm	240
	Electromagnetic						
NDI	Aurora	AC	0.9 ;0.3°	0.5 x 0.5 x 0.5	4 (6 DOF)	200 x 200 x 70	40
Ascension	Flock of Birds	Pulsed DC	1.8 RMS	1.2 x 1.2 x 1.2	4	240 x 290 x 66	50
Polhemus	FASTTRAK	AC	0.038;0.15°	1.52 x 1.52 x 1.52	4		120

As compared to optical and EM localizers, mechanical tracking systems demonstrate significantly better robustness to external influences, such as light or feromagnetic elements in the field. Optical systems are widely used for their contact-less tracking and good accuracy. Optical tracking is limited to surface (or close to surface) measurements, since a line-of-sight between the camera and rigid bodies is required. In some cases, the tip of a long rigid surgical tool can be tracked using a rigid body on the proximal end of the tool. However, the accuracy dramatically decreases with the distance from the rigid body. Electromagnetic systems can localize objects inside a body, offering a tracking solution for some minimally invasive procedures. The major limiting factor for the EM systems is their relatively low accuracy and significantly smaller working volume as compared to optical systems. Similar to optical systems, EM trackers can track multiple tools at the same time.

The challenge of registration is to find the relationship between the tracked points on the patient and tools with the corresponding locations in medical images. Methods for performing registration are described in the next section.

3. Registration Algorithms

In a typical image guided surgical setting, patient preoperative images are registered to a set of sparse points on the patient in the operating room and/or set of points of the robotic systems. In this chapter, point-based and surface-based registration techniques are described. Other registration problems, such as registration based on mutual information or 2D-3D matching are out of the scope of this chapter.

In general terms, registration is an optimization procedure establishing a transformation (mapping), T, between two reference frames:

$$T : R_m \mapsto R_d \Leftrightarrow T(\vec{m}) = \vec{d}, \tag{1}$$

where R_m and R_d are two coordinates systems (CS), measurement and data, and $\vec{m}$ and $\vec{d}$ are points in measurement and data CS, respectively. This terminology is used here since in image guidance, one set of matching points is acquired from image data (data points) and other through tracking (measurement points).

Depending on the source and the nature of matching points, the transformation T can take different forms. **Rigid-body transformation** is a linear transformation used if the deformation between two coordinate systems is negligible.

Non-rigid transformation is a global or piecewise polynomial transformation, used if deformations are significant, for example if patient data is matched to an anatomical atlas or if a tissue deformation is to be compensated. As the mathematical theory that allows tissue deformation to be characterized is a developing field of biomechanics, this chapter will be focused on the rigid-body transformation. A rigid transformation can be represented with a 3x3 rotation matrix $\mathbb{R}$ and a 3x1 translation vector $\vec{t}$:

$$T(\vec{m}) = \mathbb{R} \cdot \vec{m} + \vec{t}. \quad (2)$$

Numerous alternative rotation representations (Euler angles, quaternions, etc.) are being used in registration algorithms, depending on data representation of matching points/surfaces and behavior of different algorithms.

3.1. Point Based Registration - Orthogonal Procrustes Problem

The orthogonal Procrustes problem in linear algebra is to find an orthogonal matrix that optimally maps a given matrix $\mathbb{A}$ to a second matrix $\mathbb{B}$ [29]. If a set of measurement points $M = \{\vec{m}_0, ..., \vec{m}_{n-1}\}$ is being matched to a set of data points $D = \{\vec{d}_0, ..., \vec{d}_{n-1}\}$, the optimization problem is to minimize squared distances between the transformed measurement points and data point, i.e. fiducial registration error (FRE):

$$FRE^2 = \sum_{i=0}^{n-1} |\mathbb{R} \cdot \vec{m}_i + \vec{t} - \vec{d}_i|^2, \quad (3)$$

by finding the optimal set $\{\mathbb{R}, \vec{t}\}$. A centroid transformation can be performed to reduce the number of free parameters in the minimization term FRE^2. If respective centroids of the measurement and data points are $\vec{\bar{m}}$ and $\vec{\bar{d}}$, sets M and D can be transformed to:

$$\widehat{M} = \{\vec{\widehat{m}}_i\}, \quad \vec{\widehat{m}}_i = \vec{m}_i - \vec{\bar{m}} \quad (4)$$

$$\widehat{D} = \{\vec{\widehat{d}}_i\}, \quad \vec{\widehat{d}}_i = \vec{d}_i - \vec{\bar{d}}. \quad (5)$$

Given that the least squares optimization presumes that both points sets have the same centroid, translation between $\widehat{M}$ and $\widehat{D}$ is defined as:

$$t = \mathbb{R} \cdot \vec{\bar{m}} - \vec{\bar{d}}, \quad (6)$$

and minimization function as:

$$FRE^2 = \sum_{i=0}^{n-1} |\mathbb{R} \cdot \vec{\widehat{m}}_i - \vec{\widehat{d}}_i|^2 \tag{7}$$

If the measurement and data points are arranged in two matrices, $\mathbb{M} = [\vec{\widehat{m}}_0 ... \vec{\widehat{m}}_{n-1}]^T$ and $\mathbb{D} = [\vec{\widehat{d}}_0 ... \vec{\widehat{d}}_{n-1}]^T$, the optimization problem reduces to Orthogonal Procrustes Problem, with the correlation matrix:

$$\mathbb{K} = \mathbb{M}^T \cdot \mathbb{D} \tag{8}$$

Using the Singular Value Decomposition (SVD), the matrix $\mathbb{K}$ can be represented as:

$$\mathbb{K} = \mathbb{U} \cdot \Sigma \mathbb{V}^T, \tag{9}$$

where $\mathbb{U}$ and $\mathbb{V}$ are are orthogonal matrices and Σ is a diagonal matrix (with same dimensionality as measurement and data points). The optimal rotation matrices, minimizing trace $\mathbb{R} \cdot \mathbb{K}$ is:

$$\mathbb{R} = \mathbb{V} \cdot \Delta \cdot \mathbb{U}^T, \tag{10}$$

where

$$\Delta = diag(1, 1, det(\mathbb{V}\mathbb{U}^T)). \tag{11}$$

Various implementations of this optimization problem have been proposed (an overview and comparison is given in [9]).

3.2. Surface Based Registration

The main objective of surface based matching is the alignment of two surfaces, one being a data surface (*model* in surface matching terminology) and the other being obtained from measurement points. The optimization criteria is the same as in the Procrustes problem, to minimize the sum of squared distances between the measurement points and the data surface.

The Iterative Closest Point (ICP) [6] is the most used surface matching algorithm. The generalized version of the ICP algorithm performs matching of two surfaces, requiring a subsampling of at least one surface. The number of points that can be collected in the operating room is usually significantly smaller than number of points acquired in the imaging studies, therefore, the ICP is very suitable for surgical applications. The ICP algorithm performs: 1) detection of closest surface point, for each measurement point and 2) point based matching with surface pairs. If the measurement points and the data surface are represented as sets of points[3], $M = \{\vec{m}_0, ..., \vec{m}_{m-1}\}$ is being registered to a set of data points $D = \{\vec{d}_0, ..., \vec{d}_{n-1}\}$, $m < n$, the closest point on the data surface, from a measurement point, in j-th iteration, can be computed as:

$$d_i^j(\vec{m}_i, D) = \min_{\vec{d} \in D}(\vec{d} - T^{j-1}(\vec{m}_i)), \tag{12}$$

where the T^{j-1} is the transformation in the $(j-1)$-th iteration. After m closest points from D are found, a point based registration is performed, as described in the section 3.1.. An obvious issue arises from Eq. 12, the ICP algorithm requires initialization of the first assumed transformation T^0. Since it is a greedy algorithm, the ICP may not converge to the global maximum and is very sensitive to the initialization. For these reasons, ICP is often used as a refinement of a point based technique. An advantage of the ICP approach is that the correspondence between the point pairs is automatically detected, as compared to point based matching, where the user has to set the correspondence manually.

3.3. Registration Accuracy

A commonly used nomenclature in the registration community classifies three registration errors [11, 12]:

1. **Fiducial Localization Error (FLE)**. The error in measuring a point location. This error arises from inaccuracies in the localization system.

2. **Fiducial Registration Error (FRE)**. The error of the registration process (for example, as in Eq. 3)

3. **Target Registration Error (TRE)**. The application error of the entire process. For example, in the surgical field the TRE relates to the error in reaching the region of interest.

For assessing the clinical impact of a registration method, TRE is the most relevant error. The squared value, as well as the distribution of the TRE relative

[3]The ICP algorithm is representation independent, for simplicity, we assume cloud of points representation

to FLE has been often analyzed and characterized. In the most prominent works [11, 12] a relationship between the expected value of TRE and FLE for the orthogonal Procrustes problem is derived:

$$< TRE^2 >=< FLE^2 > (\frac{1}{n} + \frac{1}{3}\sum_{i=1}^{3}(\frac{d_i^2}{f_i^2})), \tag{13}$$

where d_i^2 is the distance of a point in which TRE is being observed from one of three principal axes of the registration points, and f_i^2 is the sum of squared distances of the registration points to the i-th principal axis.

To minimize the TRE, with a given localization system and defined FLE, the expression $\sum_{i=1}^{3}(\frac{d_i^2}{f_i^2})$ should be minimized through an appropriate selection of the registration points or the rigid body[4]. Two important facts can be deducted from Eq. 13. Firstly, minimal $TRE_{min}^2 = \frac{<FLE^2>}{n}$ is in the centroid of the registration points and is inversely proportional to number of registration points. Secondly, iso-accuracy lines/surfaces have ellipsoidal shape with radii f_1, f_2, f_3. As the configuration of registration points becomes more collinear, the mean square distance from other two orthogonal axes is increasing, which results in high inaccuracies away from the main axis. Therefore, in order to avoid low accuracy areas, registration points should be arranged in a symmetric fashion with as many points as possible, e.g. evenly distributed on a sphere. If the points are collected from a patient, a good practice would be to arrange the registration points evenly around the area of interest, i.e. to position the area of interest in the centroid of the registration points. The TRE for some point configurations will be analyzed next.

Let us assume that $d_1^2 = d_2^2 = d_3^2 = d^2$. This means that the point in which the TRE is observed has coordinates (d, d, d) in the coordinate system described by the principal axes with origin in the center of mass of the registration points. This assumption simplifies the analysis of Eq. 13. If we assume a spherical arrangement of the registration points:

$$< TRE^2 >=< FLE^2 > (\frac{1}{n} + \frac{d}{3}\sum_{i=1}^{3}(\frac{1}{f_i^2})). \tag{14}$$

[4] In optical tracking, tracking of passive tools can be understood as a point-based tracking, with measurement points being camera detected spheres and data points being characterized geometry of the rigid body

Thus, in order to optimize the selection of registration points, the term $\sum_{i=1}^{3}(\frac{1}{f_i^2})$ has to be as small as possible. The assumed configuration can be achieved if three points are arranged in an equilateral triangle, four points in a square or a tetrahedron, five points in a pentagram, etc. Three components $f_i^2, \quad i = 1..3$ of an n-sided regular polygon with a side length a can be computed as:

$$f_1^2 = R^2 \cdot \sum_{i=0}^{n-1} \sin^2 \left[\frac{2\pi}{n}(i - \frac{1}{2}) \right] = \frac{R^2}{2} \tag{15}$$

$$f_2^2 = f_1^2 = \frac{R^2}{2} \tag{16}$$

$$f_3^2 = R^2 \tag{17}$$

where $R = \frac{a}{2} \sin^{-1}(\frac{\pi}{n})$. Therefore, it is possible to compute the $< TRE >^2$ for different numbers of registration points as:

$$< TRE^2 > / < FLE^2 > = (\frac{1}{n} + \frac{d}{3}(\frac{20 \sin^2(\frac{\pi}{n})}{a^2})) \tag{18}$$

Figure 1 shows the relationship between $< TRE >^2$ and $< FRE >^2$ for n-sided regular polygons with unit side length, on an iso-surface with unit radius. The figure also demonstrates the influence of the two parts of the $< TRE >^2$ / $< FRE >^2$ relationship ($\frac{1}{n}$ and $\frac{20 \sin^2(\frac{\pi}{n})}{a^2}$). It is obvious that adding more points to the registration improves the outcome. The selection of points and the properties of the configuration (second term) yield prominent improvement, as expected. Therefore, as the number of registration points increases, a special attention should be given to the configuration, in order to significantly improve the outcome.

In practice, it is often more desirable to have a 3D arrangement of the registration points. Since there are only nine 3D figures with points arranged on a surface of a sphere (five Platonic solids and four Kepler-Poinsot polyhedra), it is impossible to derive a generalized formula for n-sided figures. In these cases, an individual analysis of the configuration should be performed in order to find an optimal solution (see [36] for more information).

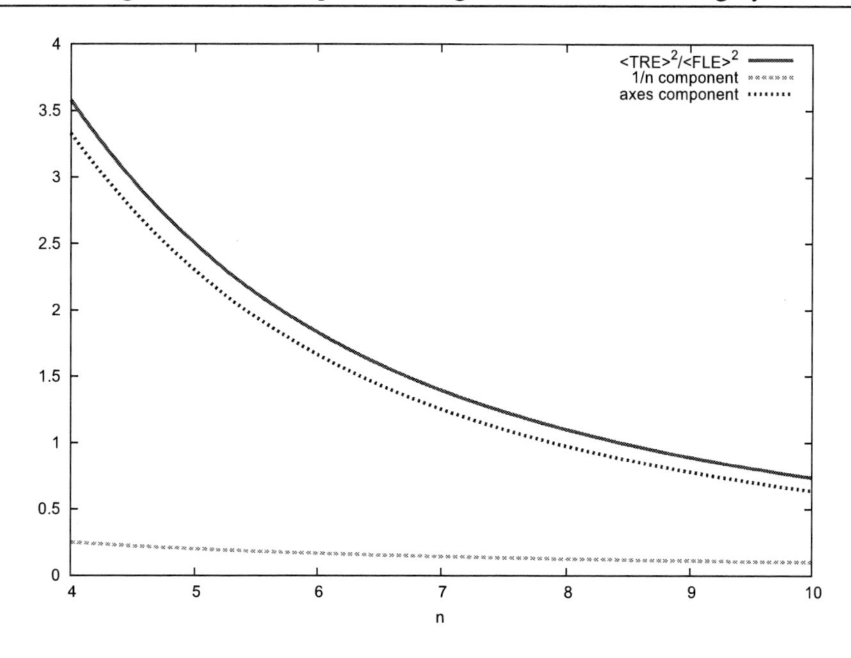

Figure 1. Relation between $< TRE >^2$ and $< FRE >^2$ for n-sided regular polygons with unit side length, on an iso-surface with unit radius

4. Registration Approaches

In this section, different approaches are desribed for registering patient images to the preoperative data as well as the robotics system to the patient.

In the operating field, there are various methods to localize the patient's geometry and acquire measurement points for a registration. Anatomical landmarks are prominent points that can be easily identified on a patient and in images. Usually, registration using anatomical landmarks is non-invasive, since it requires no artificial markers to be attached or implanted onto the patient. However in some applications, in order to access prominent points, additional excision to expose anatomy could be needed. The accuracy of registration with anatomical landmarks is also usually poor due to difficulty identifying exactly the same points in both coordinate frames. This is specially true for patient's points in the OR, given the number of possible disturbances in the environment. An alternative approach to improve registration accuracy is skin markers, easy-to-localize adhesive fiducials. However, they may still exhibit poor precision, since skin markers can shift between the imaging and intervention. An additional disadvantage is that the markers should be placed prior to imaging, and this might require special preparation procedure, e.g. shaving. The most accurate localization method for a point-based registration is the utilization of implantable markers. The markers are usually implanted in a bony structure to assure a mechanical stability. This is an invasive technique, requiring pre-imaging and additional anesthesia and causing discomfort in some patients. However, they are often used since the accuracy is significantly better compared to other techniques. Finally, surface-based techniques can be used, in which pre-registration is done using one of the above described non-invasive techniques followed by surface palpation and surface-based matching.

Registration of the robotics system to the patient and the preoperative data depends on the type and purpose of the robot, and design considerations. If a robotic system has a teleoperated master-slave architecture, as described in section 1., or in case of the passive endoscope holders, there is usually no need to explicitly register the robot to the operating field, since an operating surgeon is controlling the robotic motion directly (e.g. daVinci [3], ZEUS [28], AESOP [14], EndoAssist [1], NeuroArm [31]). Active and interactive systems usually require a three-fold registration: registration between the robotic system and the patient, registration between the robotic system and the preoperative data, and registration between the patient and the preoperative data. It is obvious that only two explicit registrations should be performed, as any of the transformations can be computed from the other two. There are two main approaches for the registration process (Figure 2):

- **Direct** registration uses the robotic system as a mechanical tracking device (see section 2.) to touch the registration points on the patient. Using this method, digitized patient points can be registered to preoperative images in the reference frame of the robot. Therefore, the full registration circle is obtained in a single procedure.

- **Indirect** registration refers to a two-phase procedure in which the relationship between a patient and preoperative data is established in a robot-independent procedure. In the image guidance techniques, for example, an optically tracked stylus touches the points. Then an explicit registration of robotic and patient reference frames (for example, registering optical rigid bodies mounted on the robot and the patient) is performed.

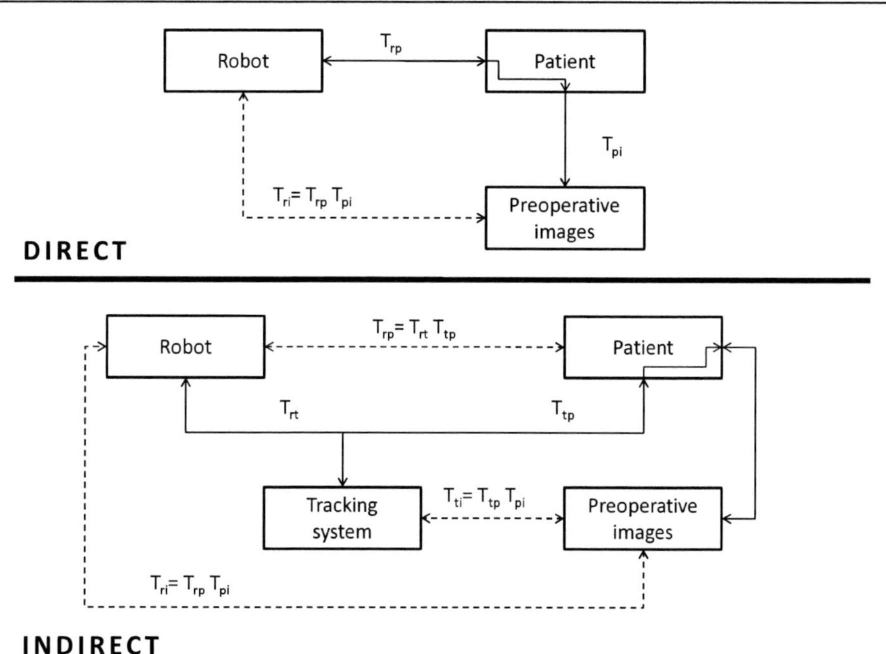

Figure 2. Conceptual overview of direct and indirect registration approaches

The direct approach is attractive for its procedural simplicity and the reduction of equipment needed for an operation, by avoiding additional tracking hardware however it encounters problems while tracking patient motion during surgery. In the indirect technique, if a dynamic reference base connected to the patient is moved together with the patient, the first registration phase is not disturbed and the robot can easily be re-registered. Alternatively, if a directly registered patient is moved, the entire direct registration process must be repeated. This might be impossible if the registration points are corrupted, such as if the markers are removed or skin incised. In an overview of the different approaches from the literature, examples of solutions to these problems are given.

4.1. Examples

CyberKnife (Accuray Inc., CA, USA) is the only FDA approved active robotic platform for radiosurgery [19, 23]. The motion of the robotic arm is determined from the planning system. Direct registration is obtained using 2D/3D image registration between two orthogonal intraoperative X-ray images (X-ray sources mounted on the ceiling) and the preoperative CT dataset. The same technique is used to track patient motion intraoperatively. The reported application error of CyberKnife is below 1 mm, whereas the error of patient tracking was 0.3 mm, 0.3° [13].

Acrobot (Acrobot, Precision Surgical Systems, London, UK) is an interactive hands-on robotic system for orthopedic surgery, particularly for unicondylar arthroplasty [2, 5, 7]. This is a special case of a master-slave architecture, with the master system residing directly on the slave part, where the robotic system imposes active constraints on the master motions performed by the surgeon. In order to transfer the preoperative CT-based planning to the intraoperative field, the Acrobot system uses a direct registration technique. For the first pre-clinical trials, implantable bone markers were used, but were later discarded as too invasive for the clinical trials [5]. The direct registration process includes surface-based registration of 20-30 points using a specially designed probe attached as a robot's end-effector [5, 7]. An initial registration is acquired using four anatomical landmarks. The problem of patient movement is solved by rigidly clamping of bones. Additional features of the clamps allow intraoperative position checking and registration update. The reported error of the registration procedure is about 2 degrees in alignment [7].

CASPAR (Universal Robot Systems (URS) Ortho, Germany) was an active robotic system used in hip and knee replacement surgery [30]. Although URS Ortho filed for a liquidation in 2004 and the CASPAR robot has been removed from the market, the registration approach is desribed here for a better understanding of the different techniques. The CASPAR system used a direct registration approach utilizing two or more bone implanted pins, each containing four markers at the top, referred to as a navigational cross. The implantation was done prior to CT scan, under local or general anesthesia and the pins were typically placed above and below the designated operation site. Intraoperatively, the robotic system in passive mode was used to localize the markers, followed by point-based matching with markers automatically detected from the CT scans. The bone motion was tracked using an NDI Polaris tracking system (see section 2.). If the absolute movement of the bone exceeded a defined threshold, the robotic system was automatically turned off and the surgical procedure is continued manually. The reported application accuracies were about 1mm in each direction (overall accuracy: $\pm$ 1.1 mm and $\pm$ 1.2°; registration accuracy $\pm$ 1mm and $\pm$ 1°mean and $\pm$ 2 mm and $\pm$ 2°maximum) [8, 30]. As discussed above, serious disadvantages of this registration approach are: the requirement for the additional operation including anesthesia as well as the inability to adapt to patient motion.

ROBODOC®(ROBODOC, a Curexo Technology Company[5], CA, USA) is an active robotic system for total hip and total knee replacement surgery. It is designed to perform an automatic milling of the prosthesis bed using a path planed using a surgical planning system (ORTHODOC®, a Curexo Technology Company). In an initial ROBODOC®configuration [17,18], the registration has been performed directly, using pins implanted in the bone by a combination of force-compliant guidance and autonomous tactile search by the robot. Patient motion is monitored using a passive, 3-DOF spherical robot attached to the exposed area of the bone, capable of detecting changes from the initial pose. If the bone is moved more than 2mm for more than 1s, the milling is interrupted, and the registration is repeated [17]. In this configuration, errors of ± 0.5 mm have been reported [17]. An alternative to pin-based registration has been proposed in [33], using a 3D/2D matching of preoperative CT and intraoperative fluoroscopy images, with maximum registration errors reported between 1.2 mm to 3.6 mm [33].

SpineAssist®(Mazor Surgical Texhnologies Ltd, Israel) is an FDA approved interactive miniature surgical assistant for placement of pedicle and translaminar facet screws [21,38]. The registration procedure for SpineAssist system is a special case of indirect registration. Registration between preoperative data and the patient is established through a 2D/3D matching of preoperative images with two intraoperative fluoroscopic images. Intraoperative robot registration requires additional two fluoroscopic images of the spine and a targeting device is attached to the patient. The targeting device is automatically detected on the images and registered to the patient. The robot can be mounted on the targeting device in a limited number of configurations, therefore, there is an explicit registration between the targeting device and the robot. This technique avoids additional tracking hardware by using intraoperative images and by directly mounting the robot on the patient, solving patient motion problems. Reported image registration accuracy is in the range of 1mm with final application error (TRE) of 0-1.5 mm measured by deviation of screw position with respect to the preoperative plan. In an alternative application [16] of the same miniature parallel robotic platform for neurosurgical key-hole surgery, another variation of indi-

[5] formerly Integrated Surgical Systems, Inc.

rect registration is applied, using a laser surface scanner and a robot mounting device with a probe. The first phase of the registration uses four anatomical landmarks to establish an initial transformation, followed by an an ICP registration of a set of skin surface points intraoperatively acquired from a laser scanner and skin points from an MRI scan. The reported overall RMS error was 1 mm. In the second phase, a transformation between a custom-made 3D registration jig attached to the robot mounting base and the patient is established using ICP registration of scanned and known surfaces of the jig. Authors report TRE for four points on the surface and inside of the phantom skull: 1.74 mm (SD $\pm$ 0.97 mm) at the entry point, 1.57 mm (SD $\pm$ 1.68 mm) at the target point, and $1.608(0.588)$ for the axis orientation. As in spine surgery, the patient motion problem is solved by attaching the robot to the patient.

PathFinder (Prosurgics, London, UK) is a surgical arm for frameless brain targeting [10]. Direct registration is achieved through a micro CCD camera mounted on the final axis of the robot, able to detect skin markers on a patient. The image processing algorithm detects titanium spheres attached to a yellow background in every framegrab of the CCD camera. This technique is used for both initial registration and patient motion tracking. The mean registration accuracy is 2.7 mm (1.8 mm - 3.2 mm) [10].

NeuroMate (Integrated Surgical Systems, CA, USA) is an FDA approved interactive robotic arm designed for holding (positioning and orienting) different tools in neurosurgery [20, 34]. The robotic system operates in two configurations, frame-based and frameless. In the frame-based configuration, the robot is directly registered to the frame whereas the registration of the frame to the patient is performed using a standardized surgical procedure. In the frameless configuration, a direct ultrasound-based registration is performed using an ultrasound transducer attached to the robot and an ultrasound marker attached to the patient. Reported application accuracies are: frame-based: mean=0.86mm, SD=0.32mm,MAX=1.30mm; frameless: mean=1.95mm, SD=0.44mm, MAX=2.69mm.

Praxiteles (Praxim, Grenoble, France) is a patient-mounted interactive guidance tool for knee and hip arthroplasty [25]. Indirect registration is performed using an optical tracking system based on digitizing important patient points. The BoneMorphing (Praxim, Grenoble, France) procedure is used to reconstruct a 3D bone model from a set of statistical models. After patient to images registration, fixation pins are inserted to pre-defined points on the patient, and the robotic system is mounted to the knee.

MIRO surgical hand[6] is a multi-purpose research platform designed for different surgical applications [15, 22]. The robot has an interactive configuration built to provide a drill guide for surgical procedures, especially pedicle screw placement. The MIRO system uses an indirect registration technique with optical tracking. The patient is registered to the preoperative data using a stylus with passive markers to digitize patient points in the dynamic reference coordinate system defined by a passive rigid body. The registration between the robot and patient is achieved by tracking the position of a passive rigid body attached to to robot's end-effector and the dynamic reference base. So far, no registration accuracy is reported.

The CRANIO robot is a parallel platform for skull tumor removal [4]. An indirect registration is done using an optically tracked A-mode ultrasound probe to digitize skull points transcutaneously for an ICP-based skull surface registration. Pre-registration is done using four anatomical markers touched with an optically tracked stylus. The robotic system is registered to the patient space by using optical tracking to record the position of robot's rigid body in the coordinate system of the patient's dynamic reference base. The phantom TRE error for US-based palpation is reported as mean 1.37mm, SD=0.51mm, which is a significant improvement from skin markers (1.78mm, SD=0.39) and anatomical landmarks (2.78, SD=1.2mm). In a cadaver study, the mean TRE was 1.97mm with SD of 1.76mm [26].

5. Summary

This chapter presents an overview of the registration techniques used in state-of-the art robotic systems for surgery and intervention. The main objective of these techniques is to transfer preoperative data to the operating field, with high accuracy, precision, and efficacy.

We offered an overview of the sensing techniques used to localize patient and surgical tools. The key issues for selection of an appropriate tracking system, is the required accuracy and the tracking environment. Optical systems are widely used due to their precision, however, they require a permanent line-of-sight between the camera and sensors. EM tracking systems can provide an alternative for intracoroporeal applications, but with limited accuracy. Mechanical arms offer excellent accuracy, but are limited to measuring only one point

[6]Previous generation was called KineMedic

at the time.

Two principal registration algorithms (point and surface-based) have been described. The accuracy analysis of point-based matching has shown that the best results can be obtained if the registration points are arranged in a 2D regular polygon or a 3D regular polyhedron. Registration accuracy is indirectly proportional to the number of points, providing with a diminished return for each additional point.

Two registration approaches are defined. Direct registration uses the robot, acting as a mechanical localization arm, to digitalize points of the patient in its coordinate system. Therefore, the transformation between the preoperative data and the robot coordinated systems is explicitly established. In an indirect registration, transformations between the patient and preoperative data and the robot and the patient are measured, and the transformation between the robot and the preoperative data is calculated from the previous three. Three main issues arise in the registration: patient motion, accuracy, and time efficiency. Direct registration is usually faster, since the process is performed in a single step, but may face problems with patient motion. In an indirect configuration, patient movement can be monitored, and the robot re-registered, since the patient-preoperative data registration remains the same.

An overview of the registration approaches used by state-of-the art robotic systems is given in the final part of this chapter. From the results reported, it is obvious that registration is the main factor in overall application accuracy. Therefore, optimization of the registration procedures and algorithms, and improvements in registration techniques are critical to robotic surgery improvement.

References

[1] S. Aiono, J. M. Gilbert, B. Soin, P. A. Finlay, and A. Gordan. Controlled trial of the introduction of a robotic camera assistant (endoassist) for laparoscopic cholecystectomy. *Surg Endosc*, **16**(9):1267–1270, Sep 2002.

[2] R. Anami, M. Kanazawa, S. Nakaura, and M. Sampei. Swing up control for acrobot with compliance of high bar focused on energy interaction with each component. In *Proc. IEEE/RSJ International Conference on Intelligent Robots and Systems IROS 2007*, pages 3334–3341, Oct. 29 2007–Nov. 2 2007.

[3] Garth H Ballantyne and Fred Moll. The da vinci telerobotic surgical system: the virtual operative field and telepresence surgery. *Surg Clin North Am*, **83**(6):1293–304, vii, Dec 2003.

[4] P. Bast, A. Popovic, T. Wu, S. Heger, M. Engelhardt, W. Lauer, K. Radermacher, and K. Schmieder. Robot- and computer-assisted craniotomy: resection planning, implant modelling and robot safety. *Int J Med Robot*, **2**(2):168–178, Jun 2006.

[5] M.D. Berkemeier and R.S. Fearing. Tracking fast inverted trajectories of the underactuated acrobot. *IEEE Transactions on Robotics & Automation*, **15**(4):740–750, Aug. 1999.

[6] P.J. Besl and N.D. McKay. A method for registration of 3-d shapes. *IEEE Transactions on Pattern Analysis and Machine Intelligence*, **14**(2):239–256, 1992.

[7] B. Davies, M. Jakopec, S.J. Harris, F. Rodriguez y Baena, A. Barrett, A. Evangelidis, P. Gomes, J. Henckel, and J. Cobb. Active-constraint robotics for surgery. *IEEE Proceedings*, **94**(9):1696–1704, Sept. 2006.

[8] Jens Decking, Christoph Theis, Tobias Achenbach, Edgar Roth, Bernhard Nafe, and Anke Eckardt. Robotic total knee arthroplasty: the accuracy of ct-based component placement. *Acta Orthop Scand*, **75**(5):573–579, Oct 2004.

[9] D. W. Eggert, A. Lorusso, and R. B. Fisher. Estimating 3-d rigid body transformations: a comparison of four major algorithms. *Mach. Vision Appl.*, **9**(5-6):272–290, 1997.

[10] M. S. Eljamel. Validation of the pathfinder neurosurgical robot using a phantom. *Int J Med Robot*, **3**(4):372–377, Dec 2007.

[11] J.M. Fitzpatrick and J.B. West. The distribution of target registration error in rigid-body point-based registration. *IEEE Transactions on Medical Imaging*, **20**(9):917–927, Sept. 2001.

[12] J.M. Fitzpatrick, J.B. West, and Jr. Maurer, C.R. Predicting error in rigid-body point-based registration. *IEEE Transactions on Medical Imaging*, **17**(5):694–702, Oct. 1998.

[13] Dongshan Fu, Gopinath Kuduvalli, Vladimir Mitrovic, William Main, and Larry Thomson. Automated skull tracking for the cyberknife image-guided radiosurgery system. In *Proc. SPIE*, volume 5744, 2005.

[14] L. K. Jacobs, V. Shayani, and J. M. Sackier. Determination of the learning curve of the aesop robot. *Surg Endosc*, **11**(1):54–55, Jan 1997.

[15] S. Jorg, M. Nickl, and G. Hirzinger. Flexible signal-oriented hardware abstraction for rapid prototyping of robotic systems. In *Proc. IEEE/RSJ International Conference on Intelligent Robots and Systems*, pages 3755–3760, Oct. 2006.

[16] L. Joskowicz, R. Shamir, M. Freiman, M. Shoham, E. Zehavi, F. Umansky, and Y. Shoshan. Image-guided system with miniature robot for precise positioning and targeting in keyhole neurosurgery. *Comput Aided Surg*, **11**(4):181–193, Jul 2006.

[17] P. Kazanzides, B.D. Mittelstadt, B.L. Musits, W.L. Bargar, J.F. Zuhars, B. Williamson, P.W. Cain, and E.J. Carbone. An integrated system for cementless hip replacement. *IEEE Engineering in Medicine and Biology*, **14**(3):307–313, May–June 1995.

[18] P. Kazanzides, J. Zuhars, B. Mittelstadt, B. Williamson, P. Cain, F. Smith, L. Rose, and B. Musits. Architecture of a surgical robot. In *Proc. IEEE International Conference on Systems, Man and Cybernetics*, pages 1624–1629, 18–21 Oct. 1992.

[19] Christopher R King, Joerg Lehmann, John R Adler, and Jenny Hai. Cyberknife radiotherapy for localized prostate cancer: rationale and technical feasibility. *Technol Cancer Res Treat*, **2**(1):25–30, Feb 2003.

[20] Qing Hang Li, Luca Zamorano, Abhilash Pandya, Ramiro Perez, Jianxing Gong, and Fernando Diaz. The application accuracy of the neuromate robot–a quantitative comparison with frameless and frame-based surgical localization systems. *Comput Aided Surg*, **7**(2):90–98, 2002.

[21] Isador H Lieberman, Daisuke Togawa, Mark M Kayanja, Mary K Reinhardt, Alon Friedlander, Nachshon Knoller, and Edward C Benzel. Bone-mounted miniature robotic guidance for pedicle screw and translaminar facet screw placement: Part i–technical development and a test case result. *Neurosurgery*, **59**(3):641–50; discussion 641–50, Sep 2006.

[22] T. Ortmaier, H. Weiss, U. Hagn, M. Grebenstein, M. Nickl, A. Albu-Schaffer, C. Ott, S. Jorg, R. Konietschke, Luc Le-Tien, and G. Hirzinger. A hands-on-robot for accurate placement of pedicle screws. In *Proc. IEEE International Conference on Robotics and Automation ICRA 2006*, pages 4179–4186, May 15–19, 2006.

[23] Todd Pawlicki, Cristian Cotrutz, and Christopher King. Prostate cancer therapy with stereotactic body radiation therapy. *Front Radiat Ther Oncol*, **40**:395–406, 2007.

[24] Graeme Penney. 2d-3d registration via digitally reconstructed radiograph (drr). *IEEE Transactions on Medical Imaging*, **17**:586–595, 1998.

[25] C. Plaskos, P. Cinquin, S. Lavalle, and A. J. Hodgson. Praxiteles: a miniature bone-mounted robot for minimal access total knee arthroplasty. *Int J Med Robot*, **1**(4):67–79, Dec 2005.

[26] Aleksandra Popovic, Stefan Heger, Axel Follmann, ting wu, Martin engelhardt, Kirsten Schmieder, and K. Klaus Radermacher. *Medical Robotics*, chapter Efficient Non-Invasive Registration with A-mode Ultrasound in Skull Surgery, pages 323–340. I-Tech Education and Publishing, 2008.

[27] Marc Rea, Donald McRobbie, Haytham Elhawary, Zion Tsz Ho Tse, Michael Lamperth, and Ian Young. System for 3-d real-time tracking of mri-compatible devices by image processing. *IEEE/ASME Transactions on Mechatronics*, **13**(3):379–382, 2008.

[28] H. Reichenspurner, R. J. Damiano, M. Mack, D. H. Boehm, H. Gulbins, C. Detter, B. Meiser, R. Ellgass, and B. Reichart. Use of the voice-controlled and computer-assisted surgical system zeus for endoscopic coronary artery bypass grafting. *J Thorac Cardiovasc Surg*, **118**(1):11–16, Jul 1999.

[29] P. H. Schonemann. A generalized solution of the orthogonal procrustes problem. *Psychometrika*, **31**:1–10, 1996.

[30] Werner Siebert, Sabine Mai, Rudolf Kober, and Peter F Heeckt. Technique and first clinical results of robot-assisted total knee replacement. *Knee*, **9**(3):173–180, Sep 2002.

[31] Garnette R Sutherland, Isabelle Latour, and Alexander D Greer. Integrating an image-guided robot with intraoperative mri: a review of the design and construction of neuroarm. *IEEE Eng Med Biol Mag*, **27**(3):59–65, 2008.

[32] F. Tatar, J.Millinger, R. C. Den Dulk, W. A. van Duyl, J. Goosen, and A. Bossche. Ultrasonic sensor system for measuring position and orientation of laproscopic instruments in minimal invasive surgery. In *Proc. 2nd Annu. Int. IEEE-EMBS Special Topic Conf. Microtechnology Medicine and Biology*, pages 301–304, 2002.

[33] R. H. Taylor, L. Joskowicz, B. Williamson, A. Guziec, A. Kalvin, P. Kazanzides, R. Van Vorhis, J. Yao, R. Kumar, A. Bzostek, A. Sahay, M. Brner, and A. Lahmer. Computer-integrated revision total hip replacement surgery: concept and preliminary results. *Medical Image Analalysis*, **3**(3):301–319, Sep 1999.

[34] T. R K Varma and P. Eldridge. Use of the neuromate stereotactic robot in a frameless mode for functional neurosurgery. *Int J Med Robot*, **2**(2):107–113, Jun 2006.

[35] S. Weiss, P. Vernickel, E. Spuentrup, M. Katoh, B. Gleich, T. Schaeffter, R. W. Guenther, and A. Buecker. In-vivo active catheter tracking using an rf-safe transmission line. In *Proc Intl Soc Mag Reson Med*, volume 13, page 196, 2005.

[36] Jay B. West and Jr. Calvin R. Maurer. Designing optically tracked instruments for image-guided surgery. *IEEE Transactions on Medical Imaging*, **23**(5):533–545, 2004.

[37] Andrew D. Wiles, David G. Thompson, and Donald D. Frantz. Accuracy assessment and interpretation for optical tracking systems. In Jr.ss Galloway, Robert L., editor, *Medical Imaging 2004: Visualization, Image-Guided Procedures, and Display*, volume 5367, pages 421–432, 2004.

[38] A. Wolf and B. Jaramaz. Mbars: Mini bone attached robotic system for joint arthroplasty. In *Proc. First IEEE/RAS-EMBS International Conference on Biomedical Robotics and Biomechatronics BioRob 2006*, pages 1053–1058, February 20–22, 2006.

In: New Robotics Research ISBN 978-1-60741-093-5
Editors: E.D. Wagner et al., pp. 131-183

Chapter 7

POSITION/FORCE CONTROL AND ITS APPLICATION TO OPEN ARCHITECTURAL INDUSTRIAL ROBOTS

Fusaomi Nagata*[1], *Keigo Watanabe*[2], *Tetsuo Hase*[3], *Zenku Haga*[3], *Masaaki Omoto*[3], *Kunihiro Tsuda*[4], *Osamu Tsukamoto*[4], *Masaki Komino*[4] *and Yukihiro Kusumoto*[5]
[1]Tokyo University of Science, Yamaguchi
[2]Saga University
[3]R&D Center, Meiho Co. Ltd.
[4]ASA Systems Inc.
[5]Fukuoka Industrial Technology Center

Abstract

In this chapter, a position/force control system is first designed for industrial robots with an open architecture controller. Position and orientation of the tool attached to the tip of an industrial robot are controlled based on the model designed by a CAD system. Also, force including kinetic friction is controlled through a desired impedance model. The both manipulated variables generated from the position control system and force control system are velocity quantity in Cartesian-coordinate system, so that the hybrid control system can be easily applied to industrial robots with an open architecture controller. Next, we introduce two examples of applications being utilized in actual manufacturing process. One is the 3D robot sander which sands the free-formed surface of wooden materials. The finished wooden workpiece with curved surface is used for a part constructing a piece of artistic furniture. The other is the mold polishing robot which finishes aluminum PET bottle blow molds. Further, the application limit of articulated-type industrial robots is quantitatively evaluated through a simple static position/force measurement. Finally, we consider a novel desktop orthogonal-type robot with higher position and force resolutions to finish a smaller workpiece such as a plastic lens mold which conventional articulated-type industrial robots have not been able to deal with. The basic position/force control performance is shown, and present research progress and promising future works are introduced.

1. Introduction

Up to now, industrial robots have drastically rationalized many kinds of manufacturing processes in industrial fields. The user interface provided by the robot maker has been almost limited to so-called teaching pendant. The teaching pendant is a useful and safe tool to obtain the position and orientation at the tip of a robot along a desired trajectory, but the teaching is very complicated and time-consuming task. Especially, when the target trajectory is a free curved line, many through points must be given to acquire a smooth trajectory; the task is further not easy.

For this decade, open architectural industrial robots as shown in Fig. 1 have been produced from several industrial robot makers such as KAWASAKI Heavy Industries, Ltd., MITSUBISHI Heavy Industries, Ltd. and YASKAWA Electric Corp. and so on. Open architecture described in this article means that the servo system and kinematics of the robot are technically opened, so that various applications required in industrial fields are allowed to be planned and developed at the user side. For example, non-taught operation by using a CAD/CAM system can be considered due to the opened accurate kinematics. Also, force control strategy using a force sensor can be implemented due to the opened servo system.

In this chapter, a position/force control system is first designed for industrial robots with an open architecture controller. Position and orientation of the tool attached to the tip of an industrial robot are controlled based on the model designed by a CAD system. Also, force including kinetic friction is controlled through a desired impedance model. The both manipulated variables generated from the position control system and force control system are velocity quantity in Cartesian-coordinate system, so that the hybrid control system can be eas-

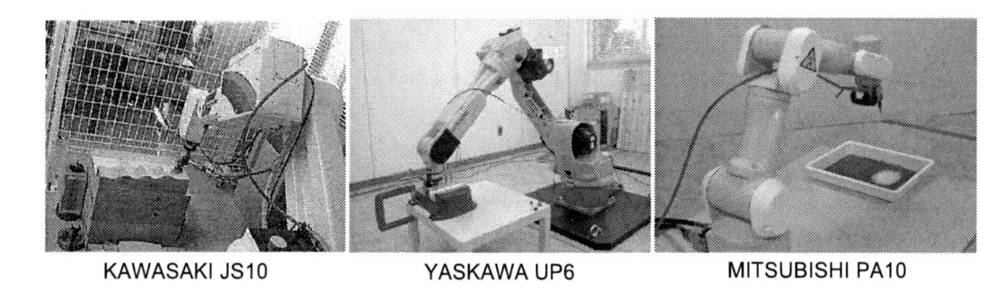

Figure 1. Industrial robots with an open architectural controller.

ily applied to industrial robots with an open architecture controller. Next, we introduce two examples of applications being utilized in actual manufacturing process. One is the 3D robot sander which sands the free-formed surface of wooden materials. The finished wooden workpiece with curved surface is used for a part constructing a piece of artistic furniture. The other is the mold polishing robot which finishes aluminum PET bottle blow molds. Further, the application limit of articulated-type industrial robots is quantitatively shown through a simple static position/force measurement. Finally, we consider a novel desktop orthogonal-type robot with higher resolutions of position and force to finish a smaller workpiece such as a plastic lens mold which conventional articulated-type industrial robots have not been able to deal with. The basic position/force control performance is shown and present research progress and promising future works are introduced.

2. Position/Force Control for Open Architectural Industrial Robots

2.1. Impedance Model Following Force/Compliance Control

Impedance control is one of the effective control strategies for a manipulator to desirably reduce or absorb the external force with an environment [1]. It is characterized by an ability which controls the mechanical impedance such as mass, damping and stiffness acting at joints. Impedance control does not have a force control mode or a position control mode but it is a combination of force and velocity. In order to control the contact force acting between the arm tip and environment, we have proposed the impedance model following force control methods (IMFFC) that can be easily implemented in industrial robots with an open architecture controller [2]. The desired impedance equation in Cartesian space for a robot manipulator is designed by

$$\boldsymbol{M}_d(\ddot{\boldsymbol{x}} - \ddot{\boldsymbol{x}}_d) + \boldsymbol{B}_d(\dot{\boldsymbol{x}} - \dot{\boldsymbol{x}}_d) + \boldsymbol{S}\boldsymbol{K}_d(\boldsymbol{x} - \boldsymbol{x}_d) = \boldsymbol{S}\boldsymbol{F} + (\boldsymbol{I} - \boldsymbol{S})\boldsymbol{K}_f(\boldsymbol{F} - \boldsymbol{F}_d) \quad (1)$$

where $\boldsymbol{x} \in \Re^3$, $\dot{\boldsymbol{x}} \in \Re^3$ and $\ddot{\boldsymbol{x}} \in \Re^3$ are the position, velocity and acceleration vectors, respectively. $\boldsymbol{M}_d \in \Re^{6\times6}$, $\boldsymbol{B}_d \in \Re^{3\times3}$ and $\boldsymbol{K}_d \in \Re^{3\times3}$ are the coefficient matrices of desired mass, damping and stiffness, respectively. $\boldsymbol{F} \in \Re^3$ is the force vector. $\boldsymbol{K}_f \in \Re^{3\times3}$ if the force feedback gain matrix. $\boldsymbol{x}_d$, $\dot{\boldsymbol{x}}_d$, $\ddot{\boldsymbol{x}}_d$ and $\boldsymbol{F}_d$ are the desired position, velocity, acceleration and force vectors, respectively. $\boldsymbol{S}$ is the switch matrix to select force control mode or compliance control mode. If $\boldsymbol{S} = \boldsymbol{0}$, Eq. (1) becomes force control mode in all directions; whereas if $\boldsymbol{S} = \boldsymbol{I}$ it becomes compliance control mode in all directions. Here, $\boldsymbol{I}$ is the identity matrix. $\boldsymbol{M}_d$, $\boldsymbol{B}_d$, $\boldsymbol{K}_d$ and $\boldsymbol{K}_f$ are set to positive-definite diagonal matrices.

When force control mode is selected in all directions, i.e., $\boldsymbol{S} = \boldsymbol{0}$, defining $\boldsymbol{X} = \dot{\boldsymbol{x}} - \dot{\boldsymbol{x}}_d$ gives

$$\dot{\boldsymbol{X}} = -\boldsymbol{M}_d^{-1}\boldsymbol{B}_d\boldsymbol{X} + \boldsymbol{M}_d^{-1}\boldsymbol{K}_f(\boldsymbol{F} - \boldsymbol{F}_d) \quad (2)$$

Here, the stability of Eq. (2) is briefly considered at the equilibrium by using Lyapunov stability analysis. The candidate of Lyapunov function [3] is proposed by

$$V(\boldsymbol{X}) = \frac{1}{2}\boldsymbol{X}^T\boldsymbol{X} \quad (3)$$

which is continuous and everywhere nonnegative. Differentiating Eq. (3) gives

$$\dot{V}(\boldsymbol{X}) = \boldsymbol{X}^T\dot{\boldsymbol{X}} = \boldsymbol{X}^T(-\boldsymbol{M}_d^{-1}\boldsymbol{B}_d\boldsymbol{X}) = -\boldsymbol{X}^T\boldsymbol{M}_d^{-1}\boldsymbol{B}_d\boldsymbol{X} \quad (4)$$

which is everywhere non-positive since $\boldsymbol{M}_d^{-1}\boldsymbol{B}_d$ is a positive definite diagonal matrix. Therefore, Eq. (3) is indeed a Lyapunov function for the system given by Eq. (2). $\dot{V}(\boldsymbol{X})$ can be zero only at $\boldsymbol{X} = \boldsymbol{0}$, everywhere else $V(\boldsymbol{X})$ decreases, so that Eq. (2) is asymptotically stable.

In general, Eq. (2) can be resolved as

$$\boldsymbol{X} = e^{-\boldsymbol{M}_d^{-1}\boldsymbol{B}_d t}\boldsymbol{X}(0) + \int_0^t e^{-\boldsymbol{M}_d^{-1}\boldsymbol{B}_d(t-\tau)}\boldsymbol{M}_d^{-1}\boldsymbol{K}_f(\boldsymbol{F} - \boldsymbol{F}_d)d\tau \quad (5)$$

In the following, we consider the form in the discrete time k using a sampling time Δt. If it is assumed that $\boldsymbol{M}_d$, $\boldsymbol{B}_d$, $\boldsymbol{K}_f$, $\boldsymbol{F}$ and $\boldsymbol{F}_d$ are constant

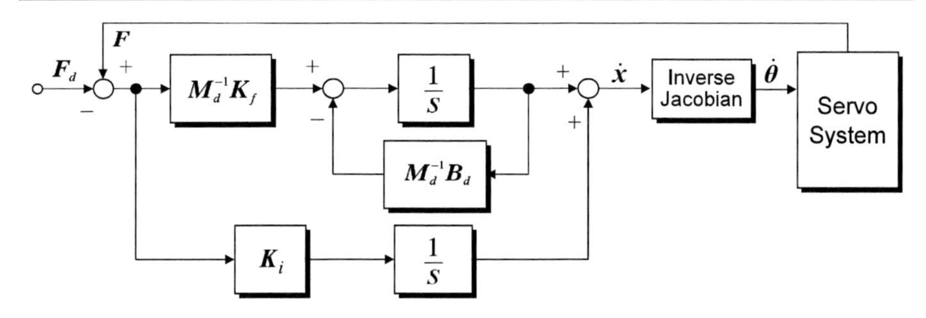

Figure 2. Block diagram of impedance model following force control with I-action.

at $\Delta t(k-1) \leq t < \Delta tk$, then defining $\boldsymbol{X}(k) = \boldsymbol{X}(t)|_{t=\Delta tk}$ leads to the recursive equation given by

$$\boldsymbol{X}(k) = e^{-\boldsymbol{M}_d^{-1}\boldsymbol{B}_d\Delta t}\,\boldsymbol{X}(k-1) - \left(e^{-\boldsymbol{M}_d^{-1}\boldsymbol{B}_d\Delta t} - \boldsymbol{I}\right)\boldsymbol{B}_d^{-1}\boldsymbol{K}_f\{\boldsymbol{F}(k) - \boldsymbol{F}_d\} \quad (6)$$

Remembering $\boldsymbol{X} = \dot{\boldsymbol{x}} - \dot{\boldsymbol{x}}_d$, giving $\dot{\boldsymbol{x}}_d = \boldsymbol{0}$ in the direction of force control, and adding an integral action, the equation of velocity command in terms of Cartesian space is derived by

$$\begin{aligned}\dot{\boldsymbol{x}}(k) &= e^{-\boldsymbol{M}_d^{-1}\boldsymbol{B}_d\Delta t}\,\dot{\boldsymbol{x}}(k-1) - \left(e^{-\boldsymbol{M}_d^{-1}\boldsymbol{B}_d\Delta t} - \boldsymbol{I}\right)\boldsymbol{B}_d^{-1}\boldsymbol{K}_f\{\boldsymbol{F}(k) - \boldsymbol{F}_d\} \\ &+ \boldsymbol{K}_i\sum_{n=1}^{k}\{\boldsymbol{F}(n) - \boldsymbol{F}_d\} \qquad (7)\end{aligned}$$

where $\boldsymbol{K}_i \in \Re^{3\times 3}$ is the integral gain matrix and is also set to a positive-definite diagonal matrix. The impedance model following force control method written by Eq. (7) is used to control the force which an industrial robot gives an environment. The block diagram of Eq. (7) is shown in Fig. 2. As can be seen, the force is regulated by a feedback control loop.

2.2. Feedfoward Position/Orientation Control

Next, we discuss position/orientation control of an end-effector attached to the tip of a robot arm. A key point is that the position/orientation control system is designed feedforwardly so as not to disturb the force feedback control loop. In almost cases, target workpieces are fortunately designed and machined

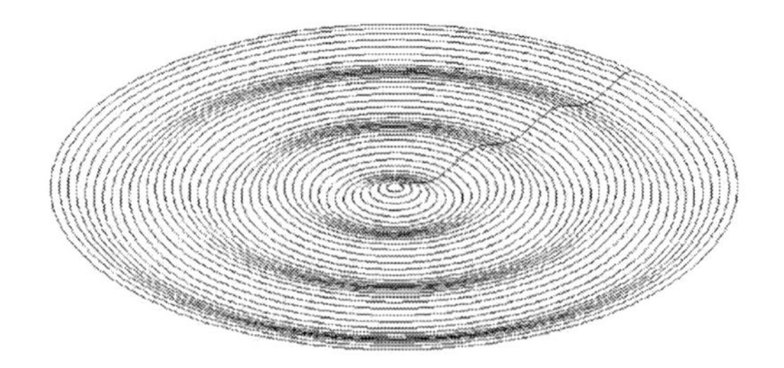

Figure 3. Example of a whirl path generated based on contour lines.

by CAD/CAM systems and NC machine tools. Therefor, cutter location data can be referred as the desired trajectory of position and orientation. The cutter location data are called CL data, and which are generated from the main-processor of CAM. The CL data are sequential points along the model surface given by a zigzag path, whirl path or spiral path. Figure 3 shows an example of a whirl path based on contour lines, which are designed by a 3D CAD/CAM Unigraphics. In order to realize non-taught operation, we have already proposed a feedforward trajectory generator [4, 5] using the CL data, which yields the desired trajectory $\boldsymbol{r}(k)$ at the discrete time k given by

$$\boldsymbol{r}(k) = \left[\boldsymbol{x}_d^T(k)\ \boldsymbol{o}_d^T(k)\right]^T \tag{8}$$

where $\boldsymbol{x}_d(k) = [x_{dx}(k)\ x_{dy}(k)\ x_{dz}(k)]^T$ and $\boldsymbol{o}_d(k) = [o_{dx}(k)\ o_{dy}(k)\ o_{dz}(k)]^T$ are the position and orientation components, respectively. $\boldsymbol{o}_d(k)$ is the normal vector at the position $\boldsymbol{x}_d(k)$. In the following, we detail how to make $\boldsymbol{r}(k)$ using the CL data.

In this approach, the desired trajectory $\boldsymbol{r}(k)$ is generated along the CL data. The CL data are usually calculated with a linear approximation along the model surface. The i-th step is written by

$$\boldsymbol{CL}(i) = [p_x(i)\ p_y(i)\ p_z(i)\ n_x(i)\ n_y(i)\ n_z(i)]^T \tag{9}$$

$$\{n_x(i)\}^2 + \{n_y(i)\}^2 + \{n_z(i)\}^2 = 1 \tag{10}$$

where $\boldsymbol{p}(i) = [p_x(i)\ p_y(i)\ p_z(i)]^T$ and $\boldsymbol{n}(i) = [n_x(i)\ n_y(i)\ n_z(i)]^T$ are position and orientation vectors, respectively. $\boldsymbol{r}(k)$ is obtained by using linear equations and a tangential velocity $\boldsymbol{v}_t(k)$ represented by

$$\boldsymbol{v}_t(k) = \left[v_{tx}(k)\ v_{ty}(k)\ v_{tz}(k)\right]^T \tag{11}$$

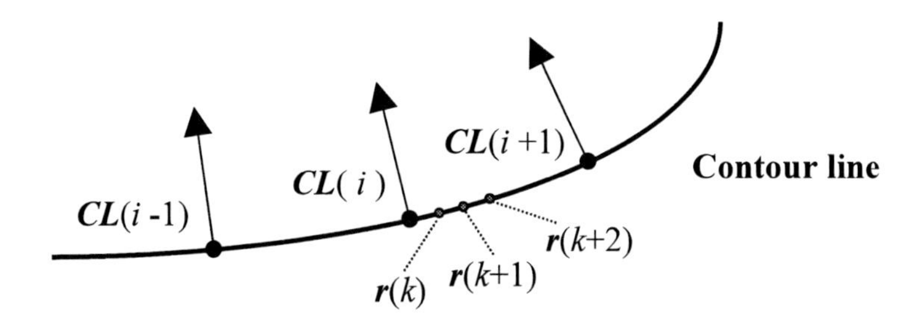

Figure 4. Relation between CL data $\boldsymbol{CL}(i)$ and desired trajectory $\boldsymbol{r}(k)$.

A relation between $\boldsymbol{CL}(i)$ and $\boldsymbol{r}(k)$ is shown in Fig. 4. In this case, assuming $\boldsymbol{r}(k) \in [\boldsymbol{CL}(i), \boldsymbol{CL}(i+1)]$ we obtain $\boldsymbol{r}(k)$ through the following procedure. First, a direction vector $\boldsymbol{t}(i)$ is given by

$$\boldsymbol{t}(i) = \boldsymbol{p}(i+1) - \boldsymbol{p}(i) \tag{12}$$

so that each component of $\boldsymbol{v}_t(k)$ is obtained by

$$v_{tj}(k) = \left\|\boldsymbol{v}_t(k)\right\| \frac{t_j(i)}{\left\|\boldsymbol{t}(i)\right\|} \ \ (j = x, y, z) \tag{13}$$

Using a sampling width Δt, each component of the desired position $\boldsymbol{x}_d(k)$ is given by

$$x_{dj}(k) = x_{dj}(k-1) + v_{tj}(k)\Delta t \ \ (j = x, y, z) \tag{14}$$

Next, the desired orientation $\boldsymbol{o}_d(k)$ is considered. We define two angles $\theta_1(i), \theta_2(i)$ as shown in Fig. 5. $\theta_1(i)$ and $\theta_2(i)$ are the tool angles of inclination and rotation, respectively. Using $\theta_1(i)$ and $\theta_2(i)$, each component of $\boldsymbol{n}(i)$ is represented by

$$n_x(i) = \sin\theta_1(i)\cos\theta_2(i) \tag{15}$$

$$n_y(i) = \sin\theta_1(i)\sin\theta_2(i) \tag{16}$$

$$n_z(i) = \cos\theta_1(i) \tag{17}$$

The desired tool angles $\theta_{r1}(k), \theta_{r2}(k)$ of inclination and rotation at the discrete time k can be calculated as

$$\theta_{rj}(k) = \theta_j(i) + \{\theta_j(i+1) - \theta_j(i)\} \frac{\left\|\boldsymbol{x}_d(k) - \boldsymbol{p}(i)\right\|}{\left\|\boldsymbol{t}(i)\right\|} \tag{18}$$

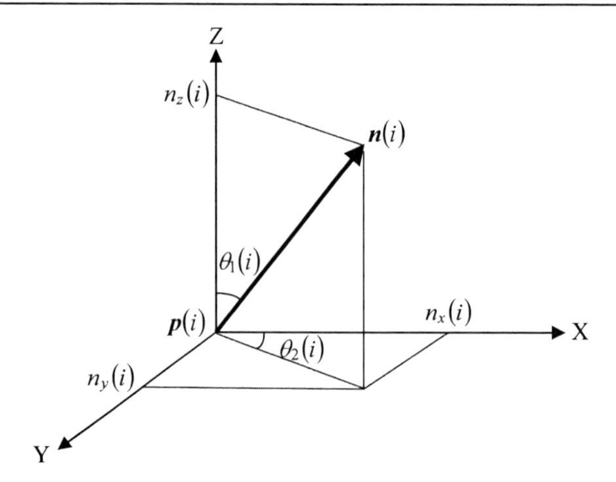

Figure 5. Normalized tool vector $\boldsymbol{n}(i)$ represented by $\theta_1(i)$ and $\theta_2(i)$ in work coordinate system.

where $j = 1, 2$. If Eq. (18) is substituted into Eqs. (15), (16), (17), we finally obtain

$$o_{dx}(k) = \sin\theta_{r1}(k)\cos\theta_{r2}(k) \tag{19}$$

$$o_{dy}(k) = \sin\theta_{r1}(k)\sin\theta_{r2}(k) \tag{20}$$

$$o_{dz}(k) = \cos\theta_{r1}(k) \tag{21}$$

$\boldsymbol{x}_d(k)$ and $\boldsymbol{o}_d(k)$ mentioned above are directly obtained from the CL data without any conventional complicated teaching, and used for the desired position and orientation of a sanding tool attached to a robot arm.

2.3. Hybrid Position/Force Control with Weak Coupling

When an industrial robot conducts a task keeping a contact with a workpiece, the position and force must be controlled simultaneously and dexterously. The hybrid position/force control method was proposed to control compliant motions of a robot manipulator, by combining force and torque information with positional data to satisfy simultaneous position and force trajectory constraints specified in a convenient task related coordinate system [6].

Here we consider a practical and applicable hybrid position/force control method which can be easily applied to open architectural industrial robots. Ba-

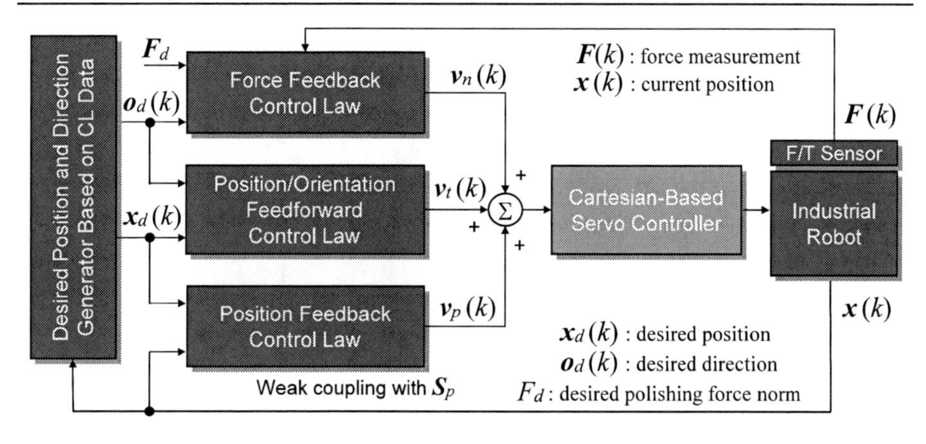

Figure 6. Block diagram of the proposed hybrid position/force controller with weak coupling.

sically, the proposed hybrid controller consists of force feedback loops and position feedforward loops in Cartesian coordinate system. Hence, the position control system and force control system do not interfere with each other, so that the position will gradually deviate from the desired trajectory due to the manipulated variable of the force feedback loop. This phenomenon is a big problem, for example, in case of developing an accurate mold polishing robot. The mold polishing robot is required to have an ability of stable force control and regular pick feed control. That means the compatibility between position control and force control should be realized. In order to cope with this problem, we have proposed a novel hybrid position/force control method with weak coupling between position feedback loop and force feedback loop. Figure 6 shows the block diagram of the proposed controller, in which position feedback loops are added only in selected directions by using a switch matrix $\boldsymbol{S}_p = [S_{px}\ S_{py}\ S_{pz}]^T$.

After this section, two applied examples of the proposed position/force control method are introduced. One is the 3D robot sander [7], the other is a mold polishing robot [8].

Figure 7. Handy air-driven sanding tools usually used by skilled workers.

3. 3D Robot Sander for Artistic Designed Furniture

3.1. Background

In manufacturing industry of wooden furniture, CAD/CAM systems and NC machine tools have been introduced widely and generally, so that the design and machining processes are rationalized drastically. However, the sanding process after machining is hardly automated yet, because it requires delicate and dexterous skills so as not to spoil the beauty and quality of the surface. Up to now, several sanding machines have been developed for wooden materials. For example, the wide belt sander is used for flat workpieces constructing furniture. Also, the profile sander is suitable for the sanding around the edge. However, these conventional machines can not be applied to the sanding task of the workpiece with free-formed surface. Accordingly, we must depend on skilled workers who can not only perform appropriate force control of sanding tools but also deal with complex curved surface. Skilled workers usually use handy air-driven tools such as a double action sanding tool and an orbital sanding tool as shown in Fig. 7.

Industrial robots have been progressed remarkably and applied to several tasks such as painting, welding, handling and so on. In these cases, it is important to precisely control the position of the end-effector attached to the tip of the robot arm. On the contrary, when the robots are applied to polishing, deburring or grinding task, it is indispensable to use some force control strategy without damaging the object. For example, polishing robots and finishing robots were presented in [9–13]. Automated robotic deburring and grinding were also introduced in [14–18].

Surface following control is a basic sanding strategy for industrial robots. It is known that two control schemes are needed to realize the surface following control system. One is the position/orientation control of the sanding tool attached to the tip of the robot arm. The other is the force control to stably keep in contact along the curved surface of the workpiece. It should be noted that if the geometric information on the workpiece is unknown, then it is so difficult to satisfactorily control the contact force moving with a high speed [19]. To suppress overshoots and oscillations, for example, the feed rate must be given a small value. Furthermore, it is also difficult to control the orientation of the sanding tool, keeping in contact with the workpiece from normal direction.

In this subsection, a robotic sanding system is integrated for new designed furniture with free-formed curved surface. The robotic sanding system provides a practical surface following control that allows industrial robots not only to adjust the polishing force through a desired impedance model in Cartesian space but also to follow a curved surface keeping contact with from normal direction. The polishing force is assumed to be the resultant force of contact force and kinetic friction force. We also describe how to apply the sanding system to a sanding task of wooden workpiece without complicated teaching process. A few sanding experiments are shown to demonstrate the effectiveness and promise of the proposed robotic sanding system using the surface following controller.

3.2. Robotic Sanding System for Wooden Parts with Curved Surface

Recently, open architectural industrial robots have been proposed to comply with user's various requests with regard to application developments. The industrial robot has an open programming interface for Windows or Linux, so that we can try to program new functions such as force control, compliance control and so on. The 6-DOF industrial robot shown in Fig. 8 is a FS30L with a PC based controller provided by Kawasaki Heavy Industries. The proposed robotic sanding system is developed based on the industrial robot whose tip has a compact force sensor. A handy sanding tool can be easily attached to the tip of the robot arm via the force sensor. A PC is connected to the PC based controller via an optical fiber cable. The PC based controller provides several Windows API (Application Programming Interface) functions, such as servo control with joint

Figure 8. Robotic sanding system developed based on an open architectural industrial robot KAWASAKI FS30L.

angles, forward/inverse kinematics, coordinate transformation and so on. By using such API functions and Windows timer, for instance, the position/orientation control at the tip of the robot arm can be realized easily and safely. In the following subsection, the surface following controller is implemented for robotic sanding by using the Windows API functions. Figure 9 shows the hardware block diagram between the controller and Windows PC. Although the standard Windows timer was set to 10 msec, the actual rate measured by an oscilloscope was sometimes about 15 msec. A stable and high sampling rate of 1 msec can be easily realized by using the Windows multimedia timer. We confirmed the sampling rate of 1 msec by using a digital data logger.

3.3. Surface Following Control for Robotic Sanding System

The robotic sanding system has two main features: one is that neither conventional complicated teaching tasks nor post-processor (CL data $\rightarrow$ NC data) is required; the other is that the polishing force acting on the sanding tool and tool position/orientation are simultaneously controlled along free-formed curved surface. In this section, a surface following control method indispensable

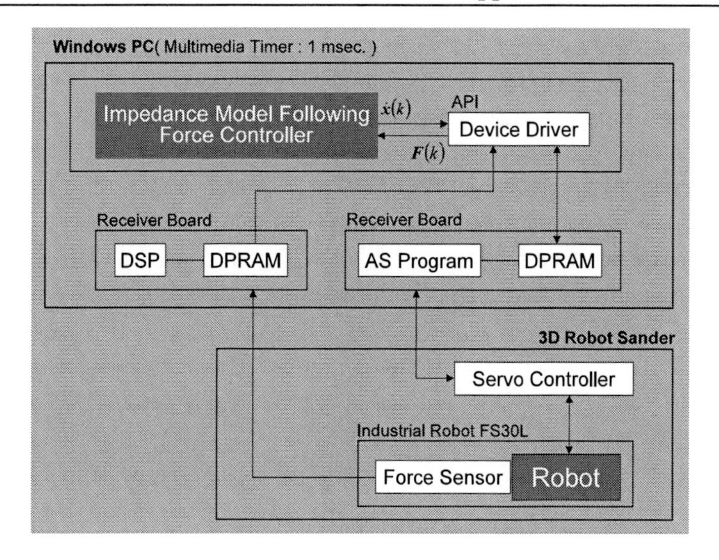

Figure 9. Hardware block diagram between Windows PC and PC controller.

for realizing the features are described in detail.

Robotic sanding task needs a desired trajectory so that the sanding tool attached to the tip of the robot arm can follow the object's surface, keeping contact with the surface from the normal direction. In executing a motion using an industrial robot, the trajectory is generally obtained in advance, e.g., through conventional robot teaching process. When the conventional teaching for an object with complex curved surface is conducted, the operator has to input a large number of teaching points along the surface. Such a teaching task is complicated and time-consuming.

Next, a sanding strategy dealing with polishing force is described in detail. The polishing force vector $\boldsymbol{F}(k) = [F_x(k)\ F_y(k)\ F_z(k)]^T$ is assumed to be the resultant force of contact force vector $\boldsymbol{f}(k) = [f_x(k)\ f_y(k)\ f_z(k)]^T$ and kinetic friction force vector $\boldsymbol{F}_r(k) = [F_{rx}(k)\ F_{ry}(k)\ F_{rz}(k)]^T$ that are given to the workpiece as shown in Fig 10, where the sanding tool is moving along on the surface from (A) to (B). $\boldsymbol{F}_r(k)$ is written by

$$\boldsymbol{F}_r(k) = -\mathrm{diag}\big(\mu_x, \mu_y, \mu_z\big)\big\|\boldsymbol{f}(k)\big\|\frac{\boldsymbol{v}_t(k)}{\big\|\boldsymbol{v}_t(k)\big\|} - \mathrm{diag}\big(\eta_x, \eta_y, \eta_z\big)\boldsymbol{v}_t(k) \quad (22)$$

where $\mathrm{diag}\big(\mu_x, \mu_y, \mu_z\big)\big\|\boldsymbol{f}(k)\big\|\big(\boldsymbol{v}_t(k)/\big\|\boldsymbol{v}_t(k)\big\|\big)$ is the Coulomb friction, and

$\mathrm{diag}(\eta_x, \eta_y, \eta_z)\boldsymbol{v}_t(k)$ is the viscous friction. μ_i and η_i $(i = x, y, z)$ are the i-directional coefficients of Coulomb friction per unit contact force and of viscous friction, respectively. Each friction force is generated by $\boldsymbol{f}(k)$ and $\boldsymbol{v}_t(k)$, respectively. $\boldsymbol{F}(k)$ is represented by

$$\boldsymbol{F}(k) = \boldsymbol{f}(k) + \boldsymbol{F}_r(k) \tag{23}$$

The polishing force magnitude can be easily measured by using a 3-DOF force sensor attached between the tip of the arm and the sanding tool, which is given by

$$\|\boldsymbol{F}(k)\| = \sqrt{\left\{{}^{S}F_x(k)\right\}^2 + \left\{{}^{S}F_y(k)\right\}^2 + \left\{{}^{S}F_z(k)\right\}^2} \tag{24}$$

where ${}^{S}F_x(k)$, ${}^{S}F_y(k)$ and ${}^{S}F_z(k)$ are the each directional component of force sensor measurements in sensor coordinate system. The force sensor used is the NITTA IFS-67M25A with a sampling rate of 8 kHz. Although the IFS-67M25A is a 6-DOF force/moment sensor, the moment components were ignored because the moment data were not needed in force control system. In the following subsection, the error $E_f(k)$ of polishing force magnitude is calculated by

$$E_f(k) = \|\boldsymbol{F}(k)\| - F_d \tag{25}$$

where F_d is a desired polishing force.

3.4. Feedback Control of Polishing Force

In the manufacturing industry of wooden furniture, skilled workers usually use handy air-driven tools to finish the surface after machining or painting. These types of tools cause high frequency and large magnitude vibrations, so that it is difficult for the skilled workers to sand the workpiece keeping the polishing force a desired value. Consequently, undesirable unevenness tends to appear on the sanded surface. In order to achieve a good surface finishing, it is fundamental and effective to stably control the polishing force. When the robotic sanding system runs, the polishing force is controlled by the impedance model following force control with integral action given by

$$\begin{aligned} v_{normal}(k) &= v_{normal}(k-1)\, e^{-\frac{B_d}{M_d}\Delta t} \\ &\quad -\left(e^{-\frac{B_d}{M_d}\Delta t} - 1\right)\frac{K_f}{B_d} E_f(k) + K_{fi}\sum_{n=1}^{k} E_f(n) \end{aligned} \tag{26}$$

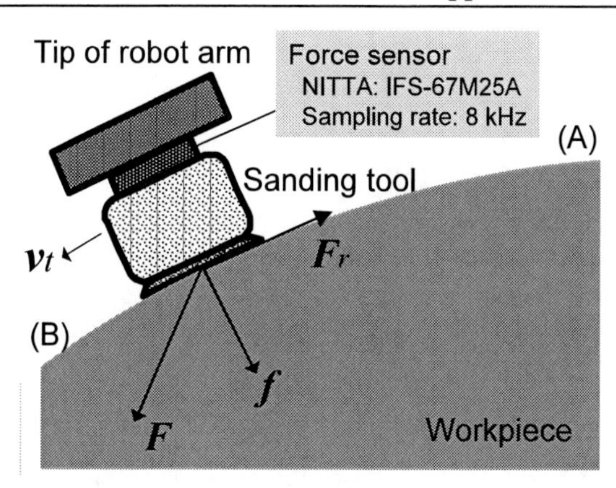

Figure 10. Polishing force $\boldsymbol{F}(k)$ composed of contact force $\boldsymbol{f}(k)$ and kinetic friction force $\boldsymbol{F}_r(k)$.

where $v_{normal}(k)$ is the velocity scalar; K_f is the force feedback gain; K_{fi} is the integral control gain; M_d and B_d are the desired mass and desired damping coefficients, respectively. Δt is the sampling width. Using $v_{normal}(k)$, the normal velocity vector $\boldsymbol{v}_n(k) = [v_{nx}(k)\ v_{ny}(k)\ v_{nz}(k)]$ at the center of the contact point is represented by

$$\boldsymbol{v}_n(k) = v_{normal}(k)\frac{\boldsymbol{o}_d(k)}{\|\boldsymbol{o}_d(k)\|} \tag{27}$$

where $\boldsymbol{o}_d(k)$ is the normal vector at the contact point, which are obtained from Eqs. (19), (20) and (21).

3.5. Feedforward and Feedback Control of Position

Currently, wooden furniture are designed and machined with 3D CAD/CAM systems and NC machine tools, respectively. Accordingly the CL data generated from the main-processor of the CAM as shown in Fig. 11 can be used for the desired trajectory of the sanding tool. The tool path (CL data) as shown in Fig 11, which are calculated in advance based on a zigzag path, is considered to be a desired trajectory of the sanding tool. Figure 6 also shows the block diagram of the surface following controller implemented in the robot sander. The position and orientation of the tool attached to the tip of the robot

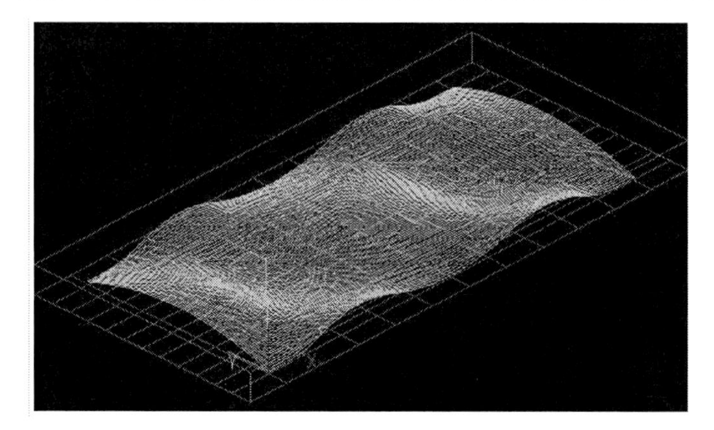

Figure 11. Zigzag path generated from main-processor of CAM. Each through point has a position and orientation component

arm are feedforwardly controlled by the tangential velocity $\boldsymbol{v}_t(k)$ and rotational velocity $\boldsymbol{v}_r(k)$, respectively referring $\boldsymbol{x}_d(k)$ and $\boldsymbol{o}_d(k)$. $\boldsymbol{v}_t(k)$ is given through an open-loop action so as not to interfere with the force feedback loop. The polishing force is regulated by $\boldsymbol{v}_n(k)$ which is perpendicular to $\boldsymbol{v}_t(k)$. $\boldsymbol{v}_n(k)$ is given to the normal direction referring the orientation vector $\boldsymbol{o}_d(k)$.
It should be noted, however, that using only $\boldsymbol{v}_t(k)$ is not enough to precisely carry out desired trajectory control along the CL data: Actual trajectory tends to deviate from the desired one, so that the constant pick feed (e.g., 20 mm) can not be performed. This undesirable phenomenon leads to the lack of uniformity on the surface. To overcome this problem, a simple position feedback loop with small gains is added as shown in Fig. 6 so that the tool does not seriously deviate from the desired pick feed. The position feedback control law generates another velocity $\boldsymbol{v}_p(k)$ given by

$$\boldsymbol{v}_p(k) = \boldsymbol{S}_p\Big\{\boldsymbol{K}_p\boldsymbol{E}_p(k) + \boldsymbol{K}_i\sum_{n=1}^{k}\boldsymbol{E}_p(n)\Big\} \tag{28}$$

where $\boldsymbol{S}_p = \text{diag}(S_{px}, S_{py}, S_{pz})$ is a switch matrix to realize a weak coupling control in each direction. If $\boldsymbol{S}_p = \text{diag}(1,1,1)$, then the coupling control is active in all directions; whereas if $\boldsymbol{S}_p = \text{diag}(0,0,0)$, then the position feedback loop does not contribute to the force feedback loop in all di-

```
FORCE/1.0
POWER/4.0
CONTACT/2.0
VELOCITY/50.0
GOTO/20.0008,300.0000,-6.8004,0.0096482,0.0893400,0.9959545
GOTO/32.1883,300.0000,-6.9395,0.0130900,0.0850632,0.9962896
GOTO/44.3758,300.0000,-7.1206,0.0165358,0.0769387,0.9968987
GOTO/56.5633,300.0000,-7.3438,0.0199903,0.0649396,0.9976890
GOTO/68.7508,300.0000,-7.6090,0.0234558,0.0490319,0.9985218
FORCE/0.000,0.000,2.000,0.000,0.000,0.000
POWER/6.0
VELOCITY/30.0
                    |
```

Figure 12. Example of proposed hyper CL data.

rections. $\boldsymbol{E}_p(k) = \boldsymbol{x}_d(k) - \boldsymbol{x}(k)$ is the position error vector. $\boldsymbol{x}(k)$ is the current position of the sanding tool attached to the tip of the arm and is obtained from the forward kinematics of the robot. $\boldsymbol{K}_p = \mathrm{diag}(K_{px}, K_{py}, K_{pz})$ and $\boldsymbol{K}_i = \mathrm{diag}(K_{ix}, K_{iy}, K_{iz})$ are the position feedback gain and its integral gain matrices, respectively. Each component of $\boldsymbol{K}_p$ and $\boldsymbol{K}_i$ must be set to small values so as not to obviously disturb the force control loop. Finally, recomposed velocities $\tilde{\boldsymbol{v}}_n(k) = [\boldsymbol{v}_n^T(k)\ 0\ 0\ 0]^T$, $\tilde{\boldsymbol{v}}_t(k) = [\boldsymbol{v}_t^T(k)\ \boldsymbol{v}_r^T(k)]^T$ and $\tilde{\boldsymbol{v}}_p(k) = [\boldsymbol{v}_p^T(k)\ 0\ 0\ 0]^T$ are summed up to a velocity command $\boldsymbol{v}(k)$, and the $\boldsymbol{v}(k)$ is given to the reference of the Cartesian-based servo controller of the industrial robot.

It can be guessed that the complete 6 constraints, which consist of 3-DOF positions and 3-DOF forces in a constraint frame, can not be simultaneously satisfied [6]. However, the delicate cooperation between the position feedback loop and force feedback loop is an important key point to successfully achieve a robotic sanding with curved surface.

3.6. Hyper CL Data

One of the features of the robot sander is that the CL data are referred as the desired trajectory of the sanding tool attached to the tip of the arm. Therefore, the complicated teaching process can be completely omitted. It is also useful that no post-processor optionally included in CAM is required to trans-

form the CL data into the NC data, i.e., the robot sander does not run based on the NC data but the CL data. The conventional CL data mainly deal with static positions and orientations along the curved surface of a model. In order to realize such a skillful sanding as skilled workers perform, we propose hyper CL data that can describe serviceable items as shown in Fig. 12. For example, following conditions can be specified:
1) Desired polishing force acting between a sanding tool and a workpiece;
2) Sanding power such as motor torque or air pressure;
3) Feed rate (tangential velocity along a curved surface);
4) Task mode change (contact mode $\leftrightarrow$ noncontact mode), etc.
The desired polishing force F_d kgf is recognized by

$$\mathrm{FORCE}/F_d \tag{29}$$

The air pressure P kgf/cm^2 of an air-driven sanding tool is regulated by

$$\mathrm{POWER}/P \tag{30}$$

The feed rate norm $\|\boldsymbol{v}_t\|$ mm/s in contact state is set by

$$\mathrm{VELOCITY}/\|\boldsymbol{v}_t\| \tag{31}$$

The task mode change from contact mode to noncontact mode is switched by

$$\mathrm{NONCONTACT}/v_m, d \tag{32}$$

where the sanding tool takes off from the workpiece with v_m mm/s to the normal direction; d mm is the distance to be moved. On the other hand, the task mode change from noncontact mode to contact mode is switched by

$$\mathrm{CONTACT}/v \tag{33}$$

where v mm/s is the approaching velocity from the normal direction. Due to the hyper CL data, it has been possible to detailedly record sanding skills.

3.7. Experimental Result

In this subsection, an experimental result of surface sanding experiment is given by using the proposed robot sander. The overview of the robot sander developed based on KAWASAKI FS20 is shown in Fig. 13. The orbital sanding

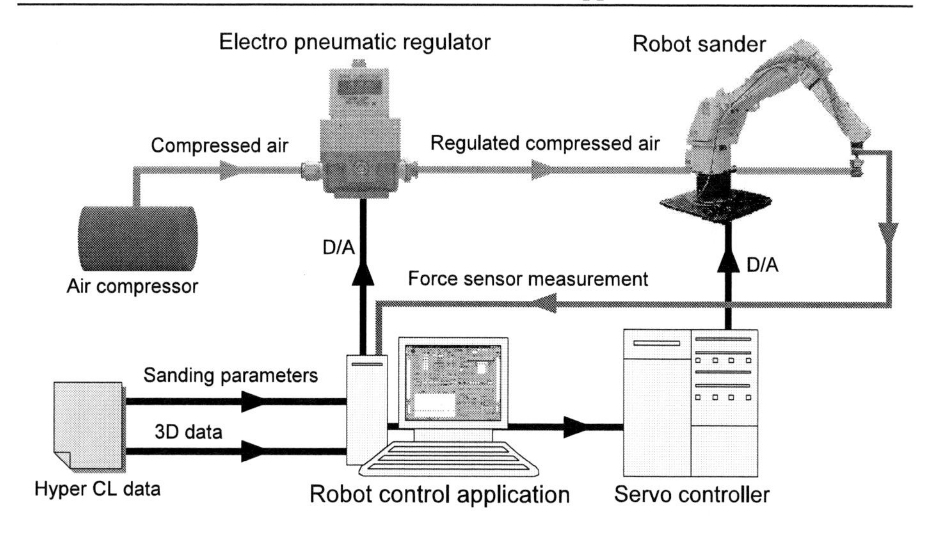

Figure 13. Overview of the robot sander developed based on KAWASAKI FS20.

tool is widely used by skilled workers to sand or finish a workpiece with curved surface. The base of the orbital sanding tool can perform eccentric motion. That is the reason that the orbital sander is not only a powerful sanding tool but also gives good surface finishing with less scratches. In this experiment, an orbital sanding tool is selected and attached to the tip of the robot arm via a force sensor. The diameter of the circular base and the eccentricity are 90 mm and 4 mm, respectively. The weight of the sanding tool is about 1.5 kg. When a sanding task is conducted, a circular pad with a sanding paper is attached to the base.

Figure 14 shows the sanding scene using the robot sander. In this case, the polishing force was satisfactorily controlled to be a desired value. Handy air-driven tools are usually used by skilled workers to sand wooden material constructing furniture. These types of tools cause large noise and vibration. Further, the system of force control consists of an industrial robot, force sensor, attachment, handy air-driven tool, zig and wooden material. Because each of them has stiff property, it is not easy to strictly keep the polishing force a constant value without overshoot and oscillation. That is the reason why measured value of the polishing force tends to have spikes and noise. However, the result would be much better than the one by skilled workers. Although it is so difficult

Table 1. Sanding conditions and control parameters

Conditions or parameters	Values
Robot	KAWASAKI FS30L
Force sensor	NITTA IFS-100M40A
Workpiece	Japanese oak
Size (mm)	$1200 \times 425 \times 85$
Diameter of sand paper (mm)	65
Grain size of sandpaper (#)	$80 \rightarrow 220 \rightarrow 400$
Desired polishing force F_d (kgf)	1.0
Feed rate $\|\boldsymbol{v}_t\|$ (mm/s)	30
Pick feed of CL data (mm)	15
Air pressure of orbital sanding tool (kgf/cm^2)	4.0
Desired mass coefficient M_d (kgf·s^2/mm)	0.01
Desired damping coefficient B_d (kgf·s/mm)	20
Force feedback gain K_f	1
Integral control gain for polishing force K_{fi}	0.001
Switch matrix for weak coupling control $\boldsymbol{S}_p$	diag(0, 1, 0)
Position feedback gain matrix $\boldsymbol{K}_p$	diag(0, 0.01, 0)
Integral control gain matrix $\boldsymbol{K}_i$ for position	diag(0, 0.0001, 0)
Sampling width Δt (ms)	0.01

and hard for skilled workers to simultaneously keep the polishing force, tool position and orientation to be desired situation even for a few minutes, the robot sander can perform the task more uniformly and perseveringly.

The left photo in Fig. 15 shows an example of the target workpieces after NC machining, i.e., before sanding, which is a representative shape that the conventional sanding machines can not sand sufficiently. The pick feed in the NC machining is set to 3 mm. The surface before sanding shown has undesirable cusp marks higher than 3 mm every pick feed. The robot sander first removed the cusp marks using a rough sanding paper ♯80, then sanded the surface using a sanding paper with a middle roughness ♯220 and finally a smooth paper ♯400. The diameters of the pad and paper were cut to 65 mm, which were so larger than that of the ball-end mill (17 mm) used in the NC machining process. The pad is put between the base and the sanding paper. Therefore, we regenerated the CL data with a pick feed 15 mm for the robotic sanding and substituted the CL data for the controller shown in Fig. 13. The contour was made so as to be a small size with an offset 15 mm to prevent the edge of the workpiece from over

Figure 14. Sanding scene using the proposed 3D robot sander developed based on a KAWASAKI FS30L.

Figure 15. Curved surfaces before and after polishing process.

sanding. Table 1 shows the other sanding conditions and the parameters of the surface following controller. These semi-optimum values were found through trial and error.

The right photo in Fig. 15 shows the surface after the sanding process. The touch feelings with both the fingers and the palm were very satisfactory. Undesirable cusp marks were not observed at all. And also, there was no over

Figure 16. Sanding scene of wooden table with curved surface.

sanding around the edge of the workpiece and no swell on the surface. Furthermore, we conducted a quantitative evaluation by using a stylus instrument, so that the measurements obtained by the arithmetical mean roughness (Ra) and max height (Ry) were around 1 μm and 3 μm, respectively. Figure 16 shows a piece of new designed artistic furniture using the workpiece sanded by the robot sander. It was confirmed from the experimental result that the proposed robotic sanding system could successfully sand the woody workpieces with curved surface.

4. Mold Polishing Robot

4.1. Background

Recently, open architectural industrial robots, whose kinematics and servo control are technically opened, have been developed. Using such a robot we can explore new skillful applications without conventional complicated teaching. In the previous section, a 3D robot sander has been introduced for the manufacturing industry of attractively designed furniture. The robot sander realized a non-taught operation regarding the position and orientation of the sanding tool attached to the robot arm. At the next stage, we try to apply an industrial robot to the polishing process of PET(Poly Ethylene Terephthalate) bottle molds. As can be guessed, the size of the target workpieces are smaller than parts constructing furniture. That means the radius of curvature is also smaller. In the manufacturing industry of PET bottle molds, also, 3D CAD/CAM systems and machining centers are being used generally and widely, and these advanced systems have drastically rationalized the design and manufacturing process of metallic molds. On the contrary, the almost polishing process after machining process has been supported by skilled workers who have capabilities concerning both dexterous force control and skillful trajectory control for an abrasive tool. The skilled workers usually use mounted abrasive tools with several sizes and shapes. In using these types of tools, keeping contact with the metallic workpiece with a desired contact force and a tangential velocity is the most important factor to obtain a high quality surface. When performing a polishing task, it is also a key point that skilled workers reciprocatingly move the abrasive tool back and forth along the object surface.

Generally, since the repetitive position accuracy at the tip of articulated-type industrial robots is 0.1 mm or its neighborhood [21], it is very difficult to polish the surface of the metallic mold using only position control strategy. In the polishing process of PET bottle molds, the surface accuracy Ra (arithmetical mean roughness) of 0.1 μm or less is finally required for mirror-like finishing. Especially, when an industrial robot contacts to a metallic workpiece, several factors that decrease the total stiffness of the system are included. They are called a clearance, strain and deflection, all of which exist in not only the robot itself but also force sensor, abrasive tool, jig, base frame and so on. Therefore, it is meaningless to discuss the position accuracy at the tip of the abrasive tool attached to the robot arm. If a position control only is used for a polishing task where an abrasive tool and metallic workpiece contact to each other, then both the stiffness of the robot itself and the total stiffness including the abrasive tool must be extremely high. However, it would be really beyond our power and also the uncertainty of workpiece positioning would not be covered.

Up to now, several robots in which force control methods are implemented have been developed; they have allowed us to achieve successful automations required in each manufacturing process. For example, polishing robots and finishing robots were presented in [22,23]. In [22], a dynamic model that describes the dynamic behavior of the robot for surface finishing tasks is developed. Also, robotic surface finishing system based on a planar robot with a force sensor and a deburring tool is proposed. In [23], the process development of a robotic system for automatically grinding and polishing vane airfoils is reported. However, each paper suggests that it is not easy to flexibly achieve the target tasks by using mere force control method. The difficulty is clear from the results that special hardware mechanisms such as manipulators, tools, sensors and measuring systems have been considered according to each manufacturing process. There also exists a serious problem that should be overcome at the present stage. When a polishing robot is designed, an end-effector with a force sensor is proposed for polishing, however, friction force acting between the end-effector and workpiece has not been successfully handled. The friction force in mold polishing has large effect on the quality and efficiency. Unfortunately, curved surface with a large curvature must be addressed in the case of polishing PET bottle molds, so that the required goal is considerably high compared to the already introduced furniture sanding robot. Additionally, when a force-controlled robot contacts with a stiff environment as a metallic mold, dynamic stability issues should be dealt with [24]. The force control system becomes unstable with the rise of the stiffness.

It is actually known that no advanced polishing robots have been successfully developed yet on a commercial basis for such metallic molds with curved surface as PET bottle molds, due to the poor polishing quality and the complicated operation. The reasons why conventional polishing robots based on an industrial robot could not satisfactorily finish the curved surface of molds are listed as follows:
1) Conventional industrial robots provide only a teaching pendant as a user-interface device. Precise teaching along a curved surface is extremely difficult and complicated.
2) Kinematics and servo control, which are indispensable to develop a real-time application for mold polishing, have not been technically opened to engineers and researchers.
3) No successful control strategy has been proposed yet for mold polishing with curved surface. Compatibility between force control and position control is need for higher surface quality. In this subsection, dexterous techniques are presented for realizing a skillful mold polishing robot as shown in Fig. 17. Normalized tool vectors are first generated from 3-axis CL data. The CL data with normal vectors called multi-axis CL data can be used for not only a desired trajectory of tool translational motion but also contact directions given to a mold. The impedance model following force control method adjusts the polishing force composed of a contact force and kinetic friction force. A CAD/CAM-based po-

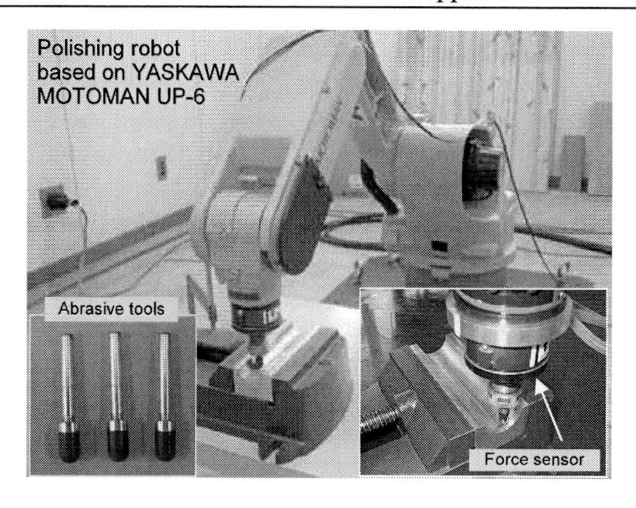

Figure 17. Mold polishing robot based on YASKAWA MOTOMAN-UP6.

sition/force controller in Cartesian space by referring such multi-axis CL data is proposed for the polishing robot with a ball-end abrasive tool. The surface polishing is achieved by controlling both the tool position along the CL data and the polishing force. The differences with the block diagram shown in Fig. 6 are that the orientation of the tool is always fixed to z-axis in work coordinate system; the orientation component in CL data are used for the force direction to be given to a mold. The CAD/CAM-based position/force controller is applied to an industrial robot with an open architectural controller. The effectiveness and validity of a mold polishing robot with the CAD/CAM-based position/force controller are demonstrated through actual polishing experiments.

4.2. Generation of Multi-axis Cutter Location Data

Although high-end 3D CAD/CAM systems generally have the multi-axis function, ordinary CAD/CAM systems do not support the function. When the 3D CAD/CAM used has no multi-axis functions, multi-axis CL data can not be generated. In this subsection, it is described that how to construct multi-axis CL data from 3-axis ones. The multi-axis CL data are composed of position and normalized tool vectors. The 3-axis CL data have position vectors $\boldsymbol{p}(i) = [p_x(i)\ p_y(i)\ p_z(i)]^T$ along the tool motion in work coordinate system, where i denotes the i-th step in CL data. Figure 18 shows the generation image of normal vector $\boldsymbol{n}(i)$ at $\boldsymbol{p}(i)$. First of all, a direction vector $\boldsymbol{t}(i) = [t_x(i)\ t_y(i)\ t_z(i)]^T$ which represents the tool moving direction is given by

$$\boldsymbol{t}(i) = \boldsymbol{p}(i+1) - \boldsymbol{p}(i) \tag{34}$$

Next, $\boldsymbol{p}(i+j)(1 \leq j \leq 100)$, which is the nearest point to $\boldsymbol{p}(i)$, is searched. Another vector $\boldsymbol{s}(i) = [s_x(i)\ s_y(i)\ s_z(i)]^T$ is obtained by

$$\boldsymbol{s}(i) = \boldsymbol{p}(i+j) - \boldsymbol{p}(i) \tag{35}$$

Note here that j is selected so that $\boldsymbol{t}(i)$ and $\boldsymbol{s}(i)$ are not parallel to each other. The normalized tool vector $\boldsymbol{n}(i) = [n_x(i)\ n_y(i)\ n_z(i)]^T$ at $\boldsymbol{p}(i)$ on the free formed surface is perpendicular to both $\boldsymbol{t}(i)$ and $\boldsymbol{s}(i)$, so that it yields the following relations:

$$n_x(i)t_x(i) + n_y t_y(i) + n_z(i)t_z(i) = 0 \tag{36}$$

$$n_x(i)s_x(i) + n_y(i)s_y(i) + n_z(i)s_z(i) = 0 \tag{37}$$

$$\{n_x(i)\}^2 + \{n_y(i)\}^2 + \{n_z(i)\}^2 = 1 \tag{38}$$

$$\gamma \geq 0 \tag{39}$$

$\boldsymbol{n}(i)$ is calculated from Eqs. (36), (37), (38) and (39). The multi-axis CL data $\boldsymbol{CL}(i) = [\boldsymbol{p}^T(i)\ \boldsymbol{n}^T(i)]^T$ $(i = 1, 2, 3, \cdots, l)$ which have information of normal direction can be composed by applying the above operations to all steps in 3-axis CL data. l is the total number of steps in CL data. When the polishing robot runs, $\boldsymbol{p}(i)$ and $\boldsymbol{n}(i)$ are used to calculate the references of position and contact direction given to curved surface, respectively. The multi-axis CL data allow the polishing robot to simply realize a non-taught operation of position and contact direction. Desired position $\boldsymbol{x}_d(k)$ and desired contact direction $\boldsymbol{o}_d(k)$ at the discrete time k are obtained similarly in subsection 2.2. The position/force controller described in next subsection refers $\boldsymbol{x}_d(k)$ and $\boldsymbol{o}_d(k)$ as desired values given to a ball-end abrasive tool, respectively.

4.3. Basic Polishing Scheme for a Ball-end Abrasive Tool

In this subsection, a control strategy efficiently using the contour of a ball-end abrasive tool is introduced for the mold polishing with curved surface. In

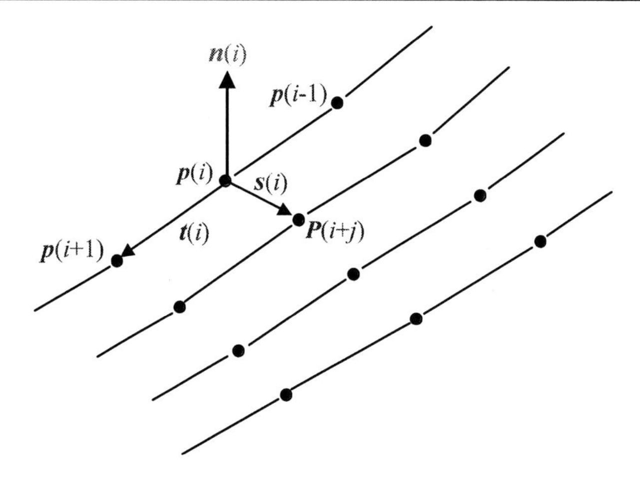

Figure 18. Generation of normalized tool vector $\boldsymbol{n}(i)$ at $\boldsymbol{p}(i)$ on a free-formed surface.

polishing, the polishing force acting between the abrasive tool and target mold is controlled. The polishing force is the most important physical factor that largely affects the quality of polishing, and assumed to be the resultant force of contact and kinetic friction forces.

The image of the proposed mold polishing robot is shown in Fig. 17, in which a ball-end abrasive tool with a radius of 5 mm is attached to the tip of a 6-DOF articulated industrial robot through a force sensor. The abrasive tool is generally attached to a portable electric sander, so that the power of polishing is obtained by its high rotational motion, e.g., 10,000 rpm. In this case, however, it is so difficult for skilled workers to suitably keep regulating the power, contact force and tangential velocity for many minutes according to object's shapes, i.e., undesirable over-polishing tends to occur frequently. Thus, to protect the mold surface against the over-polishing, the proposed polishing robot keeps the rotation of the tool fixed, and polishes the workpiece using the resultant force $\boldsymbol{F}_r$ of the Coulomb friction and the viscous friction. Each friction force is generated by the contact force $\boldsymbol{f} = [f_x\ f_y\ f_z]^T$ in normal direction and the tangential velocity $\boldsymbol{v}_t = [v_{tx}\ v_{ty}\ v_{tz}]^T$, respectively. Figure 19 shows the control strategy taking account of the kinetic friction forces. In this figure, $\boldsymbol{f}$ is given by the normal velocity $\boldsymbol{v}_n = [v_{nx}\ v_{ny}\ v_{nz}]^T$ at the contact point between the abrasive tool and mold. $\boldsymbol{v}_n$ is yielded by the IMFFC given by Eq. (7).

In this chapter, the polishing force is defined as the resultant force of $\boldsymbol{F}_r$ and $\boldsymbol{f}$, which can be measured by a force sensor. Figure 20 shows an example of force sensor, ATI Mini40 6-axis force/torque sensor. It is assumed that the polishing is performed by a hybrid control of the tool position and polishing force. In order to avoid the interference between the abrasive tool and mold, the orientation of the tool is not changed and always fixed to z-axis in work coordinate system. Fortunately, since PET bottle molds have no over-hang, a suitable contact point between the ball-end abrasive tool and the mold can be always obtained. The proposed polishing robot does not need to use any complex tools, vision sensors and jigs, so that it can be realized in a simple manner.

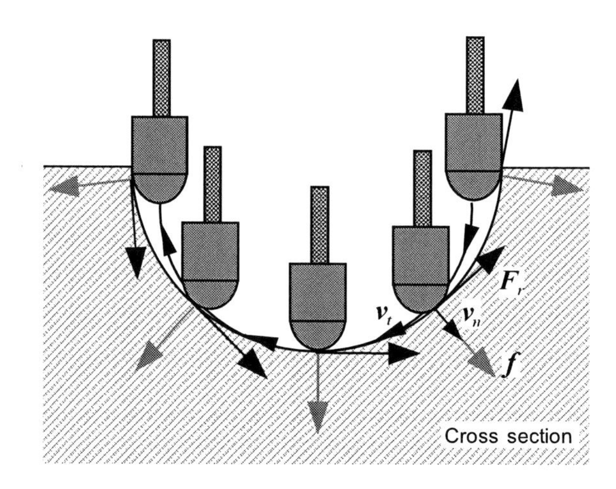

Figure 19. Polishing strategy by efficiently using the contour of a ball-end abrasive tool.

4.4. Feedback Control of Polishing Force

As mentioned above, if it is assumed in polishing that the friction force $\boldsymbol{F}_r$ acting on the abrasive tool is mainly composed of Coulomb and viscous frictions, then $\boldsymbol{F}_r$ is represented by Eq. (22). The polishing force assumed in the previous subsection is obtained by the resultant force of the contact force $\boldsymbol{f}(k)$ in normal direction and the friction force $\boldsymbol{F}_r(k)$ in tangential direction.

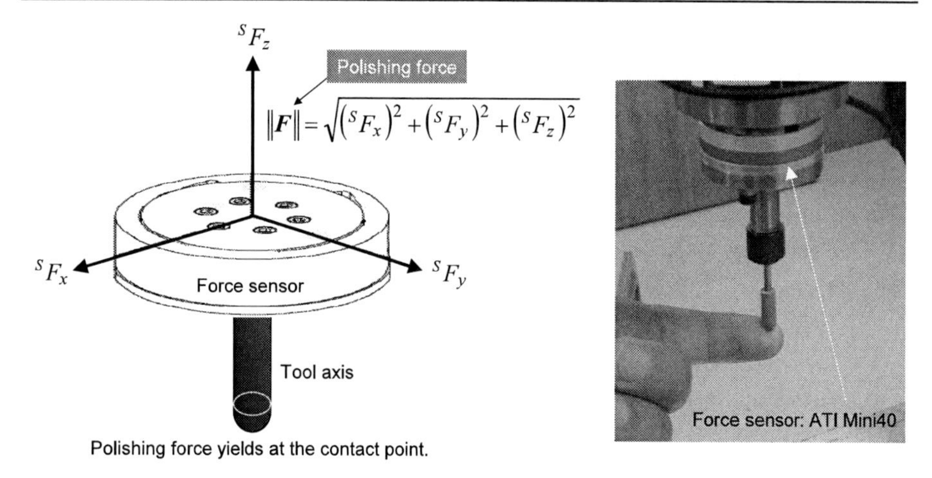

Figure 20. Force sensor to measure the polishing force $\|\boldsymbol{F}\|$.

The force vector $\boldsymbol{F}(k) = [F_x(k)\ F_y(k)\ F_z(k)]^T$ measured by the force sensor as shown in Fig. 20 is regarded as the polishing force, so that it can be also written by Eq. (23). The polishing force given by Eq. (23) is controlled by the IMMFC given by Eq. (7). In the proposed mold polishing robot, the IMFFC is applied only in normal direction at the contact point, and the control law yields a velocity $v_{normal}(k)$ given by Eq. (26). Using the $v_{normal}(k)$ and normal direction vector $\boldsymbol{o}_d(k)$ included in Eq. (8), the normal velocity vector $\boldsymbol{v}_n(k)$ to control the polishing force is obtained from Eq. (27).

4.5. Feedforward and Feedback Control of Tool Position

Currently, the molds for PET bottle manufacturing are designed and machined with 3D CAD/CAM systems and machining centers, respectively. Accordingly, multi-axis CL data or 3-axis CL data generated from the main processor of the CAM can be used for the desired trajectory of an abrasive tool. If only 3-axis CL data can be generated, normal direction vectors needed are obtained by a simple calculation described in subsection 4.2. The block diagram of the CAD/CAM-based position/force controller implemented in the polishing robot is almost same as Fig. 6, except the tool axis is always fixed to z-axis in work coordinate system. In other words, the tool orientation is not made to change in order to keep the force control stability and to uniformly abrade the contour of the ball-end abrasive tool.

The position of the abrasive tool is feedforwardly controlled by the tangential velocity $\boldsymbol{v}_t(k)$ given by

$$\boldsymbol{v}_t(k) = v_{tangent} \frac{\boldsymbol{x}_d(k) - \boldsymbol{x}_d(k-1)}{\|\boldsymbol{x}_d(k) - \boldsymbol{x}_d(k-1)\|} \tag{40}$$

where $v_{tangent}$ is a velocity norm which means the feed rate. $\boldsymbol{v}_t(k)$ is given through an open-loop action so as not to interfere with the normal velocity $\boldsymbol{v}_n(k)$. On the other hand, the polishing force is regulated by $\boldsymbol{v}_n(k)$ which is perpendicular to $\boldsymbol{v}_t(k)$. $\boldsymbol{v}_n(k)$ is given to the normal direction referring $\boldsymbol{o}_d(k)$. It should be noted, however, that using only $\boldsymbol{v}_t(k)$ is not enough to execute desired trajectory control along the CL data, i.e., the tool would not be able to conduct regular pick feed motion, e.g., with a given pick feed of 0.1 mm. To avoid this undesirable phenomenon, a simple position feedback loop with a small gain is added as shown in Fig. 6 so that the abrasive tool does not deviate from the desired pick feed motion. The position feedback control law generates another velocity $\boldsymbol{v}_p(k)$ given by Eq. (28). Finally, the velocities $\boldsymbol{v}_n(k)$, $\boldsymbol{v}_t(k)$ and $\boldsymbol{v}_p(k)$ are summed up, and which is given to the reference of the Cartesian-based servo controller of the industrial robot. The CAD/CAM-based position/force controller deals with neither the moment nor rotation, and also the origin of the constraint space (force space) is always chosen at the contact point. Accordingly, although a paper by Duffy [20] states the fallacy of modern hybrid control theory such as dimensional inconsistency, dependence on the choice of origin of the coordinates, our proposed system is not affected at all.

4.6. Experiment

4.6.1. Experimental Setup

To evaluate the validity and effectiveness of the mold polishing robot using the CAD/CAM-based position/force controller, a fundamental polishing experiment is conducted using an aluminum mold machined by a machining center as shown in Fig. 21. The main objective of the fundamental polishing is to remove all cusp marks on the curved surface whose heights are roughly 0.3 mm. The fundamental polishing before finishing is one of the most important processes to make the best beauty for mirror-like surface finishing. If the undesirable cusp marks are not uniformly removed in advance, then it is very

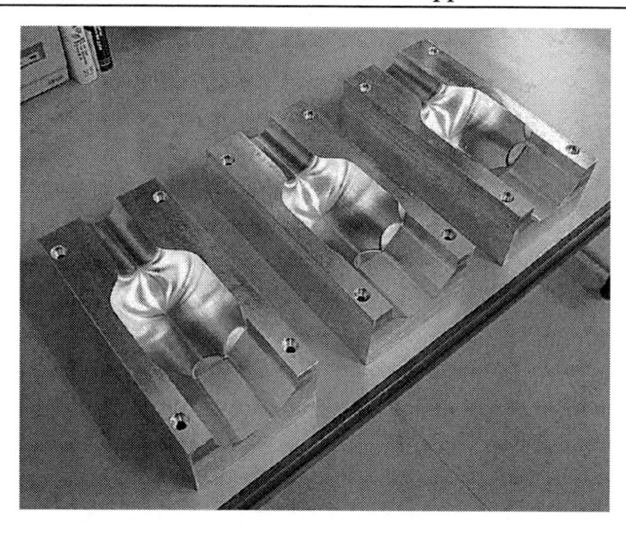

Figure 21. Target PET bottle blow mold before polishing process.

difficult to finish the mold with mirror-like surface without scratches, swells and over-polishing however much time would be spent to the finishing process.

Figure 22 shows the mold polishing robot developed based on an industrial robot YASKAWA MOTOMAN-UP6 with open control architecture used in the polishing experiment. The industrial robot provides several useful Windows API functions such as a Cartesian-based servo control and kinematics. A ball-end abrasive tool is attached to the tip of the robot arm via a force sensor. The robot also has a tiltable jig to incline the workpiece flexibly, and an oil mist to clean the surface of the mold and abrasive tool.

4.6.2. Polishing Condition and Experiment

An aluminum mold is fixed along the y-axis in robot work coordinate system with the tiltable. The mold is inclined with 45 degrees about the x-axis. One of the CL data used was generated with a zigzag path along the cross section as shown in Fig. 23(a). Mounted abrasive tools are usually attached to a portable electric sander, whose torque makes the abrasive tools rotate. When skilled workers manually use such an electrically driven tool, the contact force is given under about 10 N and its neighborhood. On the other hand, when a skilled worker polishes a mold by friction forces using a simple bamboo stick

Figure 22. Experimental setup.

covered with a sand paper, the contact force is given about from 10 to 40 N. In this experiment, since the rotation of the mounted abrasive tool is locked and the polishing task is conducted using the contact and kinetic friction forces, the desired polishing force is set to 20 N. Locking the rotation of the tool reduces undesirable high-frequency vibrational noises which are one of the serious problems in the case of allocating the force sensor between the tip of the robot arm and the abrasive tool. The force values sensed every 125 μsec were filtered with a cut off frequency of 500 Hz.

The surface was polished through three steps, making the grain size of the abrasive tool smaller gradually, i.e., from ♯220, ♯320 to ♯400. To obtain a higher quality surface, another zigzag path as shown in Fig. 23(b) was also used for the basic trajectory of the abrasive tool. These two passes shown in 23 were used alternately. Figure 24 shows the polishing scene using the proposed robot. When

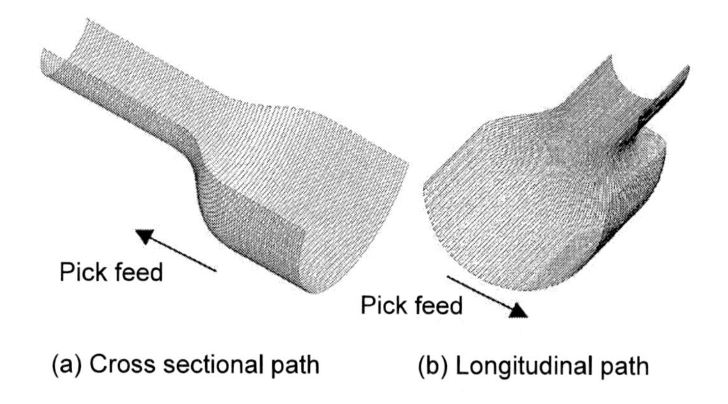

Figure 23. Desired trajectories composed of zigzag paths.

the polishing robot runs, the abrasive tool is made to reciprocatingly rotate with ± 40 deg/sec using the 6-th axis of the robot so that the contour can be abraded uniformly. If the abrasive tool is uniformly abraded keeping the ball-end shape, the robot can keep up the initial performance of polishing. Although the tool length gradually becomes shorter due to the tool abrasion, the force controller absorbs the uncertainty concerning the tool length. Other polishing conditions in the case of using the path shown in Fig. 23(a) are tabulated in Table 2. As can be seen from the values of the force and position feedback gains, the y-directional position feedback loop delicately contributes to the force feedback loop in order to keep the constant pick feed even around the inclination part shown in Fig. 24.

4.6.3. Compliant Motion of a Ball-End Abrasive Tool

The CAD/CAM-based position/force controller allows the polishing robot to perform a compliant motion so as not to damage the mold surface as illustrated in Fig. 25. The CL data shown in Fig. 25 are extracted from Fig. 23(a). First of all, the abrasive tool approaches to the initial contact point (3) with a low speed of 1 mm/s from a just above point. The position/force controller immediately becomes active when a contact occurs, i.e., the force sensor detects a small value more than 2 N in z-direction. Then, the abrasive tool starts trajectory tracking motion along the CL data. As can be seen from Fig. 25, the

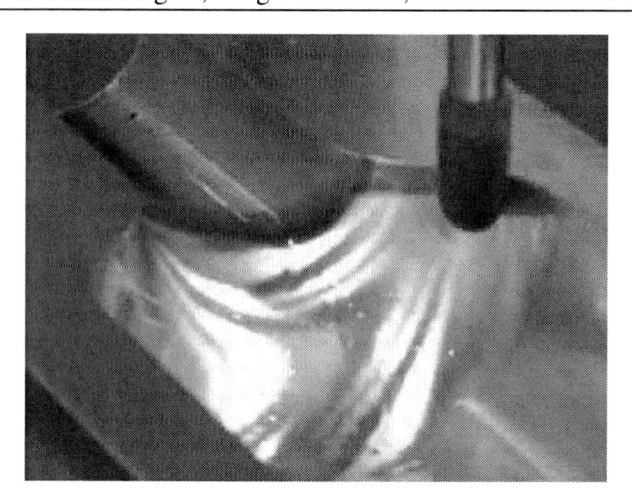

Figure 24. Experimental scene using the proposed mold polishing robot.

Table 2. Polishing conditions in the case of using the path shown in Fig. 23(a)

Conditions	Values
Robot	YASKAWA MOTOMAN UP6
Force sensor	NITTA IFS-67M25A
Zigzag path along curved surface	Cross section
Pick feed in longitudinal direction	0.2 mm
Radius of abrasive tool R	5 mm
Grit size of abrasive tool	$\sharp 220, \sharp 320, \sharp 400$
Rotational velocity of 6-th axis	40 or -40 deg/s
Rotational limits of 6-th axis	$-90 < \theta_6 < 90$ deg
Desired polishing force F_d	20 N
Tangent directional velocity $\|\boldsymbol{v}_t\|$	8 mm/s
Desired mass coefficient M_d	0.01 N·s^2/mm
Desired damping coefficient B_d	30 N·s/mm
Force feedback gain K_f	1
Force integral control gain K_i	0.0001
Position feedback gain K_{vx}, K_{vy}, K_{vz}	$0, 0.02, 0$
Sampling time Δt	10 ms

force controller absorbs position errors in x-direction caused by the radius of the abrasive tool; the higher the tool position is, the larger the position error between the CL data and actual trajectory is. If the tip of the abrasive tool stiffly moves along the CL data, then the mold or tool would be fractured. Figure 26 shows an x-directional actual trajectory acquired from an experiment in which the CL data as shown in Fig. 13 were used in the cross section and $\|\boldsymbol{v}_t\|$ was set to 6 mm/s. It is observed from this result that the abrasive tool attached to the tip of the robot arm was controlled with a desirable compliance in x-direction. Of course, such compliant motion also works well in other two directions according to the shape of mold.

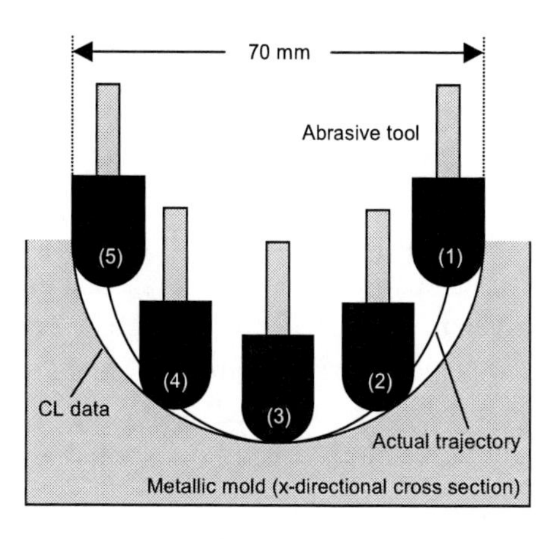

Figure 25. Comparison of CL data and actual trajectory.

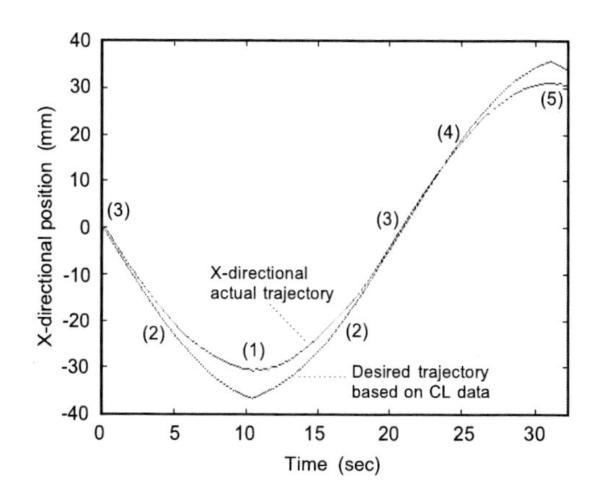

Figure 26. X-directional actual trajectory obtained through an experiment.

4.6.4. Experimental Result

It was confirmed that the significant resolution in force domain was almost about 1.2 N with a position resolution of 0.01 mm in the case that the repetitive position accuracy at the tip of the abrasive tool shown in Fig. 17 was about 0.1 mm [26]. Although the force resolution is not so fine compared with skilled workers taking charge of the mold polishing process, the polishing robot can perform each task more uniformly and stably for many hours. The surface state after polishing was evaluated by both eye checking and touch feeling with fingers, and consequently a successful surface without over-polishing around the edges was observed. Furthermore, we conducted a quantitative evaluation by using a stylus instrument, so that the measurements obtained by the arithmetical mean roughness (Ra) and maximum depth (Ry) were around 0.1 μm and 1.6 μm, respectively. It can be seen from the result that the polishing force could be more uniformly given to the surface on the average compared to skilled workers due to the proposed position/force controller. This is a main reason why such an effective polishing quality was obtained drastically. Figure 27 shows the curved surface after wiped with a cloth containing a polishing compound Cr_2O_3.

Up to now, it has been widely discussed and believed in manufacturing industries of metallic molds that it would be impossible for articulated type industrial robots, whose position accuracy is 0.1 mm at most, to polish and finish metallic molds with curved surface. However, our proposed polishing robot could satisfactorily overcome this problem due to the CAD/CAM-based position/force controller. In this section, we have proposed a CAD/CAM-based position/force controller in Cartesian space for a mold polishing robot. The CL data with normal vectors are referred as not only the desired trajectory of tool translational motion but also the desired contact direction given to molds, so that a complete non-taught operation of position and contact direction can be realized. The controller also regulates the polishing force consisting of the contact and kinetic friction forces. When the robot carries out polishing tasks, the position control loop delicately contributes to the force control loop to achieve both a regular pick feed control along the CL data and a stable polishing force control on curved surface. The effectiveness and promise of the polishing robot using the CAD/CAM-based position/force controller have been proved by actual polishing experiments using an industrial robot YASKAWA MOTOMAN-UP6.

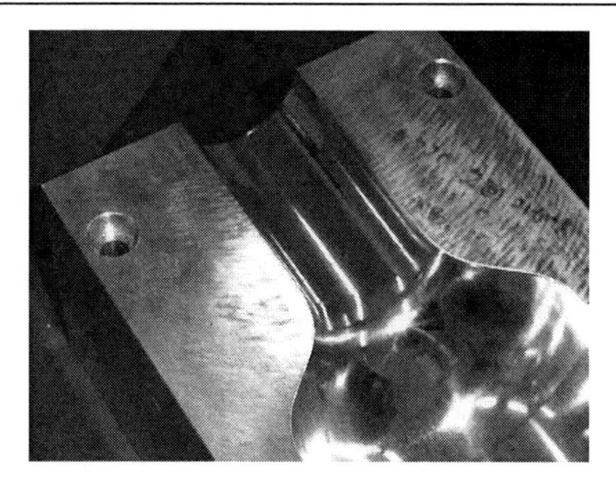

Figure 27. Constriction part of mold surface after wiped with a cloth containing the polishing compound Cr_2O_3.

5. Desktop Orthogonal-Type Robot for Smaller Workpieces

5.1. Background

The polishing of pickup lens mold after machining process requires high accuracy, delicateness and skill, so that it has not also been automated yet. Generally, a target mold has several concave areas precisely machined with a tolerance of ± 0.01 mm as shown in Fig. 28. In this case, each diameter is 4 mm. That means the target mold is not axis-symmetric, so that conventional effective polishing systems, which can deal with only axis-symmetric workpiece, can not be applied. Accordingly, such an axis-asymmetric lens mold as shown in Fig. 28 is polished by skilled workers in almost cases. Skilled workers generally polish a small lens mold by using a wood stick tool with diamond paste while checking the polished area through a microscope. However, the smaller the workpiece is, the more difficult the task is. In particular, it is required for a pickup lens mold to handle the surface uniformly and softly, so that high resolutions of position and force are indispensable. As an example of using an NC machine tool with high position resolution, Lee et al. proposed a polishing system composed of two subsystems: they are a three-axis machining center and a two-axis polishing robot. Although a sliding mode control algorithm with velocity compensation

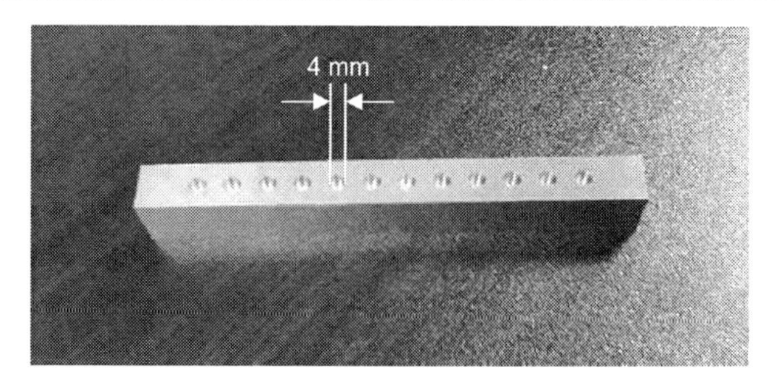

Figure 28. Pickup lens mold with several concave areas precisely machined within a tolerance, which is a representative axis-asymmetric workpiece.

was implemented to reduce tracking errors, some method which can absorb undesirable positioning error of each workpiece was not considered [25].

In this section, a new desktop orthogonal-type robot with compliance controllability is presented for finishing metallic molds with small curved surface [26]. The orthogonal-type robot consists of three single-axis robots. A tool attached to the tip of the z-axis has a ball-end shape. Also, the control system of the orthogonal-type robot is composed of a force feedback loop, position feedback loop and position feedforward loop. The force feedback loop controls the polishing force consisting of tool contact force and kinetic friction force. The position feedback loop controls the position in spiral motion direction. Further, the position feedforward loop leads the tool tip along a spiral path. It is expected that the orthogonal-type robot delicately smooths the surface roughness with about 50 μm height on each concave area, and finishes the surface with high quality.

In order to first confirm the application limit of a conventional articulated-industrial robot to a finishing task, we evaluate the backlash that causes the position inaccuracy at the tip of the abrasive tool, through a simple position/force measurement. Through a similar position/force measurement and a surface following control experiment along a lens mold, the basic position/force controllability of the proposed orthogonal-type robot is demonstrated.

Figure 29. Polishing scene of a PET bottle blow mold using a conventional articulated industrial robot with a servo spindle.

5.2. Limitation of a Polishing System Based on an Articulated-type Industrial Robot

In subthis section, effective stiffness, valid position resolution of Cartesian-based servo system and resultant force resolution in a polishing system based on an articulated-type industrial robot are evaluated. The effective stiffness means the total stiffness including the characteristics composed of an industrial robot itself, force sensor, attachment, abrasive tool, workpiece, zig and floor. Figure 29 shows the tip of the arm a finishing system developed based on a conventional articulated industrial robot YASKAWA MOTOMAN-UP6. A compact force sensor is attached to the arm tip, and a ball-end abrasive tool is fixed to the tip through a servo spindle.

Figure 30 shows the relation between the position and contact force obtained by a simple contact experiment. The position written in the horizontal axis is the z-directional component at the tip of an abrasive tool, which is calculated by the forward kinematics using the joint angles obtained from the inner sensors. The force written in vertical axis is yielded by contacting the tool tip with a aluminum workpiece and measured by a force sensor. When the tool tip contacts to the workpiece, small manipulated variables under 10 μm could not cause any effective force measurements, so that we conducted the experiment while giving the minimum resolution -10 μm in press motion and 10 μm in unpress motion.

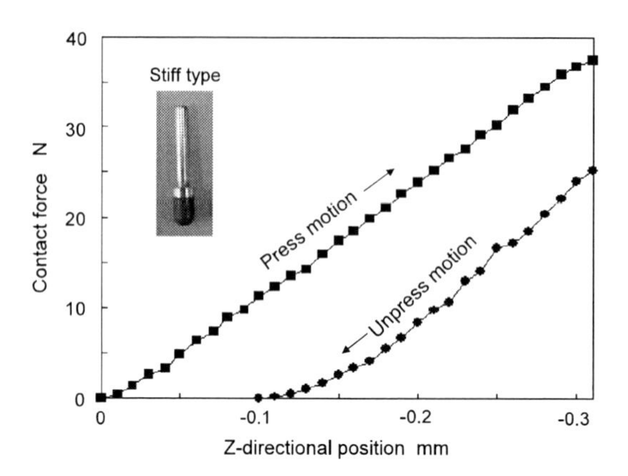

Figure 30. Static relation between position and contact force in case of a conventional articulated industrial robot.

In the experiment, the tool tip approaches to an aluminum workpiece with a low speed, and after touching the workpiece, i.e., after detecting a contact force, the tool tip was pressed against the workpiece with every 0.01 mm. The graph written with ■ in Fig. 30 shows the relation the position and contact force. The force is about 36 N when the position of the tool tip is -0.3 mm, so that the effective stiffness within the range can be estimated with 120 N/mm. After the contact motion, the tool tip was away from the workpiece once, and returned to the position again where 36 N had been obtained. After that, the tool tip was unpressed every 10 μm. The graph written with • in Fig. 30 shows the relation of the position and contact force of this case. It is observed from the result that there exists a large backlash about 0.1 mm. This value is almost the same compared to the general one that is guaranteed as a repetitive position accuracy of articulated industrial robots. That is the reason why in order to design a polishing system using an articulated-type industrial robot we must consider a force control system with the force resolution 1.2 N under the position uncertainty of 0.1 mm. Accordingly, for example, it would be very difficult for industrial robots to deal with a plastic lens mold which has the several same and small machined areas precisely arranged within the tolerance of ± 10 μm as shown in Fig. 28.

5.3. Desktop Orthogonal-type Robot with Compliance Controllability

In order to polish small concave areas as shown in Fig. 28, a novel polishing system is proposed in this subsection. Figure 31 shows the proposed desktop orthogonal-type robot consisting of three single-axis robots with a position resolution of 1 μm. The three single-axis robots are used for x-, y- and z-directional motions. A servo spindle motor is also used for the rotational motion of the tool axis. The servo spindle motor with a reduction gear and the tool axis work together with a belt. The size of the robot is 850 $\times$ 645 $\times$ 700 mm. The single-axis robot is a position control device ISPA with high-precision resolution provided by IAI Corp., which is composed of a base, linear guide, ball-screw, AC servo motor and so on. The effective strokes in x-, y- and z-directions are 400, 300 and 100 mm, respectively. The hardware block diagram is shown in Fig. 32.

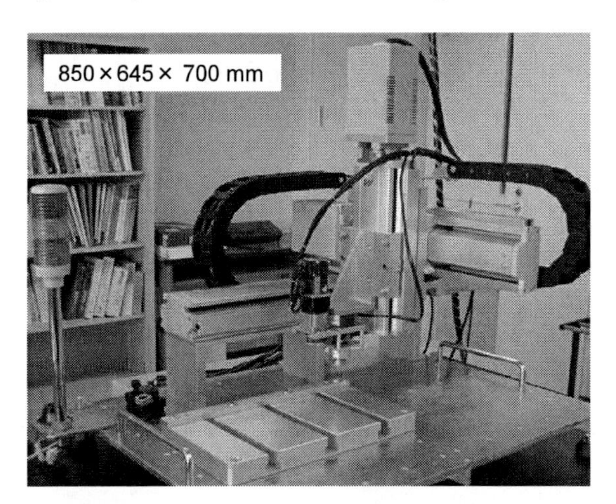

Figure 31. Proposed desktop orthogonal-type robot with compliance controllability.

Figure 33 shows the static relation between position and contact force in case of the proposed orthogonal-type robot with a stiff ball-end abrasive tool. The experiment was conducted in the same condition as shown in Fig. 30. It is observed that the undesirable backlash is largely decreased and the effective stiffness is about 30/0.2=150 N/mm. On the other hand, Figure 34 shows the relation in case that an elastic ball-end abrasive tool is used. It is observed that

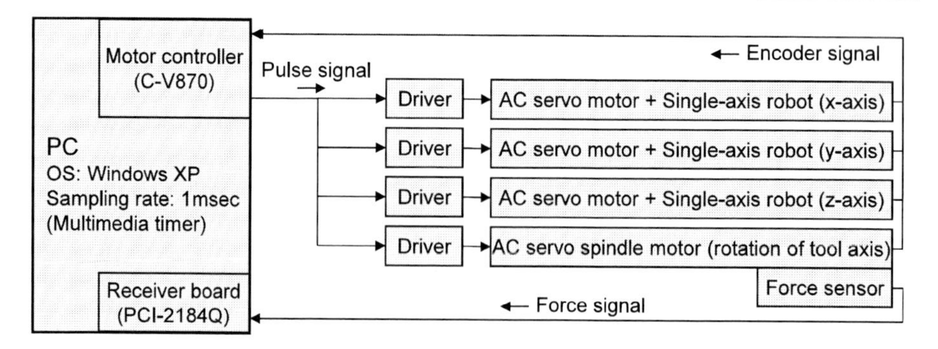

Figure 32. Hardware block diagram of the desktop orthogonal-type robot composed of three single-axis robots with a position resolution of 1μm.

the effective stiffness changes down to about 30/0.36≈83.3 N/mm according to the property of the elastic abrasive tool. As can be seen, the effective stiffness of force control system is largely affected by the kinds of the abrasive tool used. In the case of the elastic abrasive tool called a rubber abrasive tool, it is expected that the force resolution about 0.083 N can be performed due to the position resolution of 1 μm.

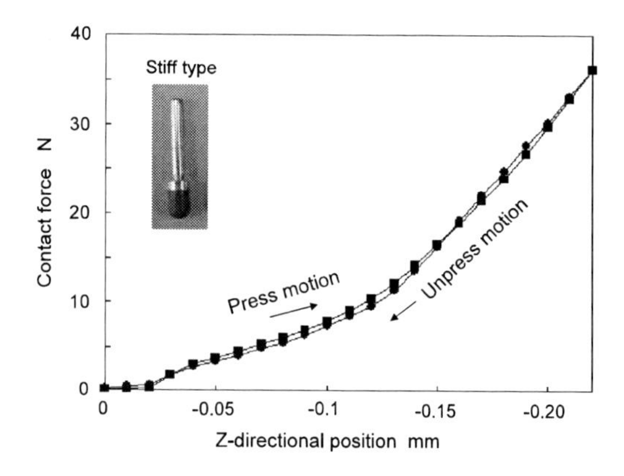

Figure 33. Static relation between position and contact force in case of the desktop orthogonal-type robot with a stiff ball-end abrasive tool.

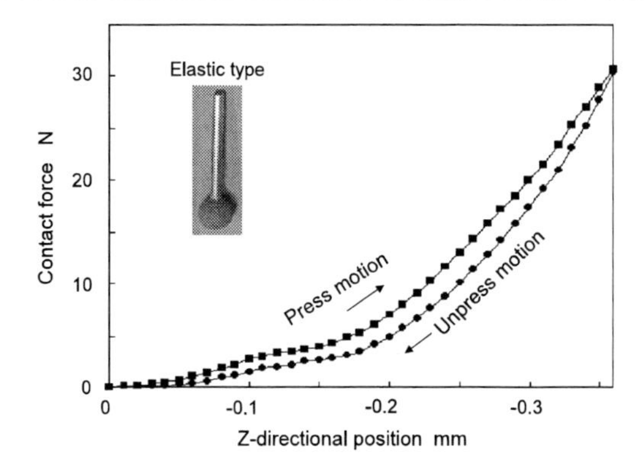

Figure 34. Static relation between position and contact force in case of the desktop orthogonal-type robot with a rubber ball-end abrasive tool.

5.4. Desired Damping Considering Critically Damped Condition

The tool tip is basically controlled by similar to the mold polishing robot in the previous section. Next, a tuning method of desired damping is proposed by using the effective stiffness of the robot. When the polishing force is controlled, the characteristics of force control system can be varied according to the combination of impedance parameters such as desired mass and damping. In order to increase the force control stability the desired damping, which has much influence on force control stability, should be tuned suitably. In this subsection, a desired damping tuning method using the effective stiffness of the robot is proposed based on the critical damping condition. When the force control mode is selected in a direction, the desired impedance model is simply written by

$$M_d(\ddot{x} - \ddot{x}_d) + B_d(\dot{x} - \dot{x}_d) = K_f(F - F_d) \tag{41}$$

where $\ddot{x}$, $\dot{x}$ and F are the acceleration, velocity and force scalars in the direction of force control, respectively. $\ddot{x}_d$, $\dot{x}_d$ and F_d are the desired acceleration, velocity and force, respectively. When the force control is active, $\ddot{x}_d$ and $\dot{x}_d$ are set to zero. It is assumed that F is the external force given by the environment and is model as

$$F = -B_m\dot{x} - K_m x \tag{42}$$

where B_m and K_m are the viscosity and stiffness coefficients of the environment, respectively. Eqs. (41) and (42) lead to the following second order lag system.

$$\ddot{x} + \frac{B_d - K_f B_m}{M_d}\dot{x} + \frac{K_f K_m}{M_d}x = 0 \tag{43}$$

The characteristics equation of Eq. (43) is written by

$$s^2 + \frac{B_d - K_f B_m}{M_d}s + \frac{K_f K_m}{M_d} = 0 \tag{44}$$

Further, solving Eq. (44) for B_d using the critical damping condition, the following condition is obtained.

$$B_d = 2\sqrt{M_d K_f K_m} - K_f B_m \tag{45}$$

In profiling control experiment, the base value for the desired damping is calculated with Eq. (45). If possible, the desired damping should be fine-tuned and varied around the base value because K_m has nonlinear characteristics as shown in Fig. 33 or 34.

5.5. Basic Experiment

In this subsection, the fundamental performance of the proposed desktop orthogonal-type robot is evaluated with respect to force control and position control. A profiling control experiment is conducted by using a plastic lens mold as shown in Fig. 35. The target mold is precisely machined based on the forms as shown in Fig. 35, whose diameter and depth under the partition line are 30 mm and 5 mm, respectively. At the start of the experiment, an elastic small ball-end abrasive tool approaches to the center of machined part with 2 mm/s. After detecting 5 N, which is 1/2 of the desired polishing force 10 N, the tool tip starts to follow a spiral path. Control parameters in the profiling control are tabulated in Table 3. The effective stiffness K_m is estimated 30/0.36$\approx$83.3 N/mm from Fig. 34 in case of the elastic ball-end abrasive tool. The desired mass M_d and force feedback gain K_f are set to 0.01, respectively. Also, it is assumed that $K_f B_m \approx 0$ since the viscosity of the force control system is considerably small. Accordingly, Eq. (45) leads to B_d=0.183 N·s/mm.

Next, control systems in x-,y- and z-directions are explained. The polishing force and z-directional position are regulated by feedback control laws, and also x- and y-directional positions are feedforwardly controlled based on CL data. The CL data, that are desired trajectory, forms a spiral path with 0.8 mm pitch

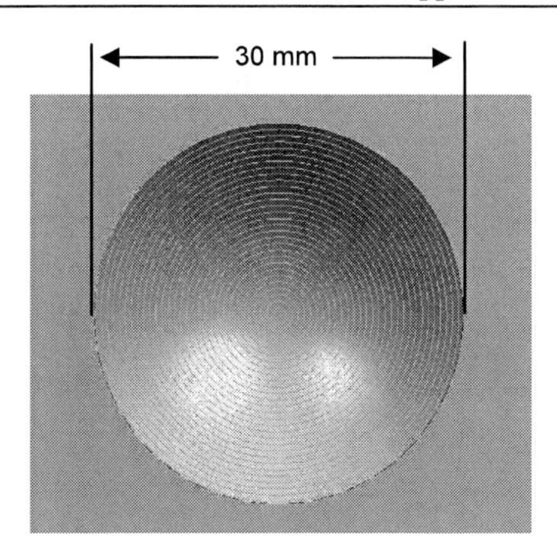

Figure 35. 3D model designed for the profiling control experiment.

Table 3. Parameters tuned for profiling control experiment.

Conditions	Values
Robot	Orthogonal-type robot
Force sensor	NITTA IFS-67M25A
Desired trajectory along curved surface	Spiral path
Pitch in X−Y plane	0.8 mm
Radius of ball-end abrasive tool R	5 mm
Grain size of abrasive tool	♯220
Desired polishing force F_d	10 N
Tangent directional velocity $\|\|\boldsymbol{v}_t\|\|$	5 mm/s
Desired mass coefficient M_d	0.01 N·s^2/mm
Desired damping coefficient B_d	0.183 N·s/mm
Force feedback gain K_f	0.01
Diagonal elements of switch matrix $\boldsymbol{S}_p$	$0, 0, 1$
Position feedback gain K_{px}, K_{py}, K_{pz}	$0, 0, 0.001$
Integral control gain K_{ix}, K_{iy}, K_{iz}	$0, 0, 0.00001$
Sampling time Δt	1 ms

viewed in x-y plane. Figure 36 shows the profiling control scene. Also, Fig. 37 shows the position control result, in which the pitch in x-y plane at the height 4 mm becomes about 0.7 mm. This is caused by that the manipulated variable

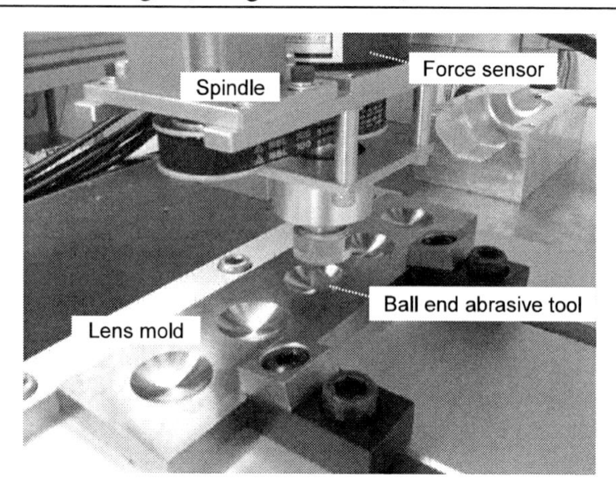

Figure 36. Profiling control scene along a spiral path.

from the force feedback control law gradually yields in x- and y- directions when the tool goes up along the spiral path. In other words, the reason is that only z-direction is designed with a position closed loop, so that the positions in x- and y-directions are corrected by the manipulated variable of the force control system, i.e., by the constraint of the force control system. Also, z-directional position was designed with a weak closed loop by using small PI gains, so that a weak coupling was conducted against to the force feedback loop. In spite of such a situation, the mean error of z-directional position in the profiling motion was about 2.7 μm. The mean error of z-directional position was evaluated by

$$E_{Pz} = \frac{\sum_{n=1}^{N} \sqrt{[x_{dz}(n) - x_z(k)]^2}}{N} \tag{46}$$

where N is the total step number in CL data, i.e., the number of 'GOTO' statement. $x_{dz}(n)$ is the z-directional component at the n step in CL data, $x_z(k)$ is the z-directional position at the tool tip obtained from an encoder. k is the discrete time when the $x_{dz}(n)$ is set.

Figure 38 illustrates the control result of the polishing force. The upper, second and third figures show x-, y- and z-directional forces in sensor coordinate system, respectively. The lower figure shows the norm $\|{}^S\boldsymbol{F}\| = \sqrt{({}^SF_x)^2 + ({}^SF_y)^2 + ({}^SF_z)^2}$. It is observed from the ${}^SF_x$ and ${}^SF_y$ in Fig. 38

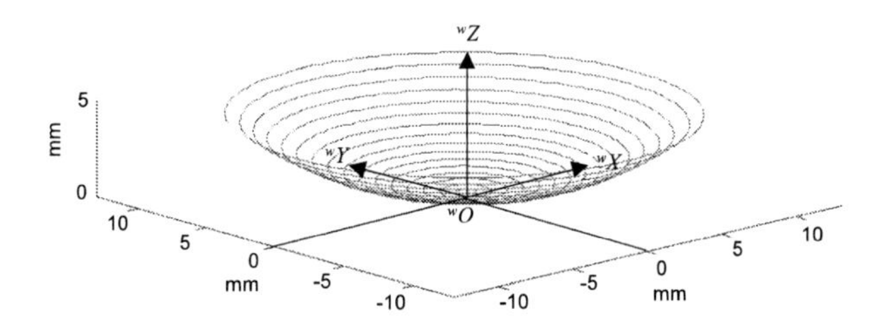

Figure 37. Position control result along the spiral path.

that the direction of force control periodically varies due to the position control along the spiral path. Also, although the force measurements in x- and y-directions increase with the rise of the tool and the z-directional force one decreased gradually, it can be confirmed that the polishing force $\|{}^{S}\boldsymbol{F}\|$ could satisfactorily follow the reference value 10 N. The cut-off frequency of the force sensor was set to 500 Hz. Figure 39 shows the normal velocity $v_{normal}(k)$ which the impedance model following force controller generated according to the force error $E_f(k)$ in the profiling control experiment. The values in Fig. 39 mean relative position commands μm per sampling time of 1 ms.

5.6. Results and Future Works

In this section, we have first examined the resolution of position and force, and effective stiffness through an experiment using an automatic polishing system based on an articulated-type 6-DOF industrial robot. Also, technical points to be improved have been considered to develop a new finishing system which can be applied to small workpieces such as a pickup lens mold. Next, an orthogonal-type robot with a static position resolution of 1 μm and force resolution of 0.083 N has been designed by combining single-axis robots. A hybrid position and force controller with compliance controllability has been also applied to the robot, in which position control, force control or their weak coupling control can coexist together. Further, we have introduced a systematic tuning method of the desired damping. The desired damping is calculated from the critically damped condition using the static relation between the position and force. Finally, a profiling control experiment using a plastic lens mold with axis symmetric surface has been conducted along a spiral path to evaluate the characteristics of position and force control. Consequently, it has been confirmed

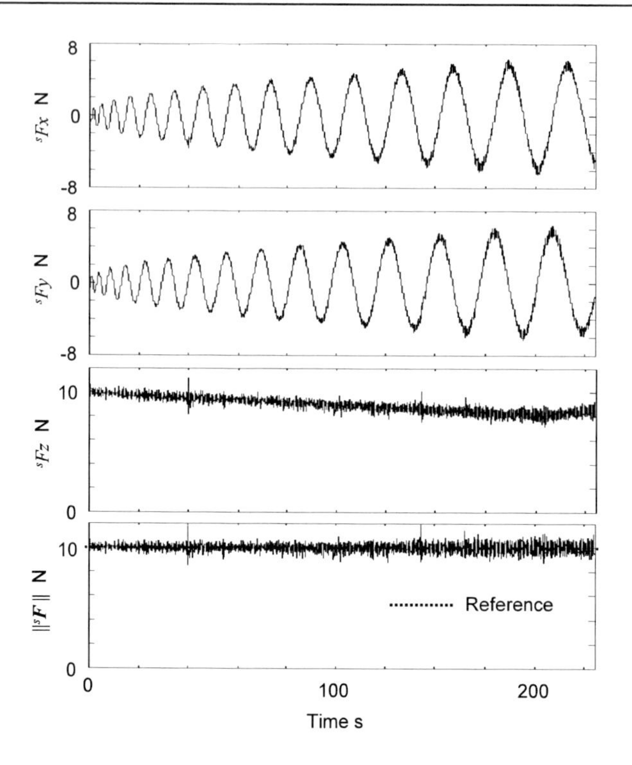

Figure 38. Control result of the polishing force $\|^S\boldsymbol{F}(k)\|$.

that the proposed orthogonal-type robot has a desirable control performance of position and force which would be applied to the finishing task of plastic lens molds.

In future work, we plan to develop a much smaller ball-end abrasive tool for finishing the workpieces as shown in Fig. 28 and to do a polishing experiment by using the proposed orthogonal-type robot with the compliance controllability.

6. Conclusion

In this chapter, a novel and practical position/force control strategy has been designed for industrial robots with an open architectural controller. Three types

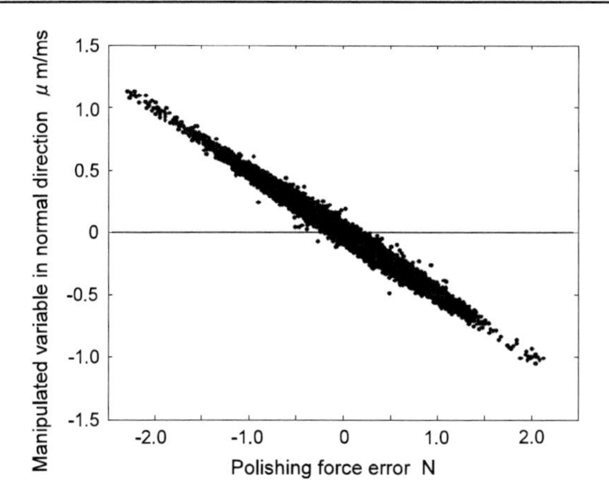

Figure 39. Relation between the polishing force error $E_f(k)$ and manipulated variable $v_{normal}(k)$ calculated from Eq. (26).

of polishing robots listed as below have been also introduced to show the effectiveness and promise.

- 3D robot sander developed based on KAWASAKI FS30L, which sands wooden parts constructing furniture with free-formed surface.
- Mold polishing robot developed based on YASKAWA MOTOMAN-UP6, which polishes aluminum PET bottle blow molds with curved surface.
- Orthogonal-type robot developed base on three single-axis robots ISPA provided by IAI Corp., which is now expected to be applied to smaller workpieces such as pickup lens molds in the future.

At the present stage, the goal of our study is the development of an orthogonal-type industrial robot with compliance controllability which can be flexibly applied to from the cusp mark removing process to the final finishing process for smaller workpieces such as pickup lens molds. In such a case, there exist two key points. One is a smaller ball-end abrasive tool with the radius of about 0.3 mm. The other is the delicate, skillful and balanced control of position and force which is ordinarily performed by skilled workers.

References

[1] Hogan, N. *Trans. ASME, J. Dyn. Syst. Measure. Contr.*, 1985, vol. 107, pp. 1–24, Impedance control: An approach to manipulation: Part I - Part III.

[2] Nagata, F., Watanabe, K., et al. *Procs. of the IEEE Int. Conf. on Industrial Electronics, Control and Instrumentation*, 2000, pp. 632-637, Polishing robot using a joystick controlled teaching system.

[3] Craig, J. J. *Introduction to ROBOTICS —Mechanics and Control Second Edition—*, Reading MA: Addison Wesley Publishing Co., 1989.

[4] Nagata, F., Watanabe, K., Sato, K. and Izumi, K. *Procs. of IEEE Int. Conf. on Systems, Man, and Cybernetics*, 1999, pp. 848–853, An experiment on profiling task with impedance controlled manipulator using cutter location data.

[5] Nagata, F., Watanabe, K. and Izumi, K. *Procs. of 2001 IEEE Int. Conf. on Robotics and Automation*, 2001, pp. 319–324, Furniture polishing robot using a trajectory generator based on cutter location data.

[6] Raibert, M. H., Craig, J. J. *Trans. ASME, J. Dyn. Syst. Measure. Contr*, 1981, Vol. 102, pp. 126–133, Hybrid position/force control of manipulators.

[7] Nagata, F., Kusumoto, Y., Fujimoto, Y. and Watanabe, K. *Robotics and Computer-Integrated Manufacturing*, 2007, vol. 23, no. 4, pp. 371-379, Robotic sanding system for new designed furniture with free-formed surface.

[8] Nagata, F., Hase, T., Haga, Z., Omoto, M. and Watanabe, K. *Mechatronics*, 2007, vol. 17, pp. 207-216, CAD/CAM-based position/force controller for a mold polishing robot.

[9] Ozaki, F., Jinno, M., Yoshimi, T., Tatsuno, K., Takahashi, M., Kanda, M., et al. *Journal of Robotics and Mechatronics*, 1995, vol. 7, no. 5, pp. 383–388, A force controlled finishing robot system with a task-directed robot language.

[10] Pfeiffer, F., Bremer, H., Figueiredo, J. *European Journal of Mechanics, A/Solids*, 1996, vol. 15, no. 1, pp. 137–153, Surface polishing with flexible link manipulators.

[11] Takeuchi, Y., Ge, D., Asakawa, N. *Procs. of IEEE Int. Conf. Robotics and Automation*, 1993, pp. 950–956, Automated polishing process with a human-like dexterous robot.

[12] Pagilla, P. R., Yu, B. *Trans. ASME, J. Dyn. Syst. Measure. Contr.*, 2001, vol. 123, pp. 93–102, Robotic surface finishing processes: modeling, control, and experiments.

[13] Huang, H., Zhou, L., Chen, X. Q., Gong, Z. M. *Int. J. of Advanced Manufacturing Technology*, 2003, vol. 21, no. 4, pp. 275–283, SMART robotic system for 3D profile turbine vane airfoil repair.

[14] Stephien, T. M., Sweet, L. M., Good, M. C., Tomizuka, M. *IEEE J. Robotics and Automation*, 1987, vol. 3, no. 1, pp. 7–18, Control of tool/workpiece contact force with application to robotic deburring.

[15] Kazerooni, H. *J. Manufacturing Systems*, 1988, vol. 7, no. 4, pp. 329–338, Robotic deburring of two-dimensional parts with unknown geometry.

[16] Her, M. G., Kazerooni, H. *Trans. ASME, J. Dyn. Syst. Measure. Contr.*, 1991, vol. 113, pp. 60–66, Automated robotic deburring of parts using compliance control.

[17] Bone, G. M., Elbestawi, M. A., Lingarkar, R., Liu, L. *Trans. ASME, J. Dyn. Syst. Measure. Contr.*, 1991, vol. 113, pp. 395–400, Force control of robotic deburring.

[18] Gorinevsky, D. M., Formalsky, A. M., Schneider, A. Yu. *Force Control of Robotics Systems*, New York: CRC Press, 1997.

[19] Takahashi, K., Aoyagi, S., Takano, M. *Procs. of the 4th Japan-France Congress & 2nd Asia-Europe Congress on Mechatronics*, 1998, pp. 398–401, Study on a fast profiling task of a robot with force control using feed-forward of predicted contact position data.

[20] Duffy, J. *Journal of Robotic Systems*, 1990, vol. 7, no. 2, pp. 139–144, The Fallacy of modern hybrid control theory that is based on orthogonal complements of twist and wrench spaces.

[21] Nagata, F., Hase, T., Haga, Z., Omoto. M. and Watanabe, K. *Procs. of the 13th International Symposium on Artificial Life and Robotics*, 2008, pp. 779–782, Intelligent desktop NC machine tool with compliance control capability.

[22] Pagilla, P. R. and Yu, B. *Trans. ASME, J. Dyn. Syst. Measure. Contr.*, 2001, vol. 123, pp. 93–102, Robotic surface finishing processes: modeling, control, and experiments.

[23] Huang, H., Zhou, L., Chen, X. Q., Gong, Z. M. *Int. J. of Advanced Manufacturing Technology*, 2003, vol. 21, no. 4, pp. 275–283, SMART robotic system for 3D profile turbine vane airfoil repair.

[24] An, H. C. and Hollerbach, J. M. *Procs. of IEEE Int. Conf. Robotics and Automation*, 1987, pp. 890–896, Dynamic stability issues in force control of manipulatorss.

[25] Lee, M. C., Go, S. J., Lee, M. H., Jun,C. S., Kim, D. S., Cha, K. D. and Ahn, J. H. *Robotics and Computer-Integrated Manufacturing*, 2001, vol. 17, no. 1–2, pp. 177–183, A robust trajectory tracking control of a polishing robot system based on CAM data.

[26] Nagata, F., Hase, T., Haga, Z., Omoto, M. and Watanabe, K. *Procs. of the 13th International Symposium on Artificial Life and Robotics*, 2008, pp. 779-782, Intelligent desktop NC machine tool with compliance control capability.

INDEX

C

D

H

I

J

K

L

M

N

O

P

Q

R

S

T

U

V

W

X

Y